Whitetails

Whitetails

NATURE'S WILD SPIRITS

Mark Raycroft

KEY PORTER BOOKS

Canadian Cataloguing in Publication Data

Raycroft, Mark
Whitetails: nature's wild spirits

ISBN 1-55263-068-4

1. White-tailed deer. I. Title.

QL737.U55R38 1999 599.65′2 C98-933000-1

The publisher gratefully acknowledges the assistance of the Canada Council and the Ontario Arts Council.

Canada

We acknowledge the financial support of the Government of Canada through the Book Publishing Industry Development Program (BPIDP) for our publishing activities.

Key Porter Books Limited
70 The Esplanade
Toronto, Ontario
Canada M5E 1R2

www.keyporter.com

Electronic Formatting: Heidi Palfrey
Design: Peter Maher

Printed and bound in Spain

99 00 01 02 6 5 4 3 2 1

I dedicate this book to my children, Martha and Andrew.
May you grow up to be strong and free,
and always able to follow your dreams.

Acknowledgments

There are many people to whom I wish to extend heartfelt thanks.

My parents, for introducing me to white-tailed deer, and for always encouraging me to pursue my dreams.

Bob McCaw, for guiding me into the world of professional wildlife photography. You are a great friend and mentor.

Bob and Alma Avery, and David Oathout, for allowing me to spend time photographing on their wonderful estate. You are kind, wonderful people. I always look forward to seeing you.

One of my best friends, Dolf DeJong, for his assistance on many of my photographic journeys. They have all been fun—and I look forward to many more.

The many dedicated people at Key Porter Books, especially Mary Ann McCutcheon, Derek Weiler, Clare McKeon, Debby de Groot, and, of course, Anna Porter.

All the deer from Alberta, Minnesota, Ontario, and New York that have allowed me to spend time with them, photographing their secret lives. It has been an honor and I will always treasure those memories.

Most importantly, I would like to thank my family—Martha, Andrew, and especially my wife, Pili—for tolerating my long hours and my many trips over the past several years. Pili, you have been my strongest supporter, always understanding and sharing my vision.

Contents

INTRODUCTION

A Frosty Morning

This book is a huge part of my heart. What I want to share here are experiences that have meant a great deal to me.

I want people to feel something of what I have felt out in nature on a good morning. Here's one of my favorite memories.

It had been one of those nights when I slept so soundly that it felt like I had just lain down when I was awakened by the relentless beeping of the alarm clock. Rubbing my eyes, I sat up and peered out the window to check the weather. The stars were still visible in the cloudless sky. The sun was below the horizon, but the blue pre-dawn light was starting to give detail to the heavily frosted meadow and trees outside the cabin. I pulled myself out of my sleeping bag and dressed, ate a hasty, probably insufficient breakfast, and mounted my camera on the hefty tripod. Five minutes later I was outside walking through the meadow. The sun was just rising over the horizon, casting rays of light between the trees. The meadow was cool and still. I headed toward a rise by a lowland clearing, where I would have a vantage point to watch for deer. Once there, my eyes scanned the clearing and the forest edge. Nothing moved. I slowly walked a short distance to a second vantage point, which faced east over the same lowland swale and thickets. After a moment something caught my eye, a movement in one of the thickets. Squinting, I saw that it was a buck rubbing his antlers on one of the trees in the thicket. Shouldering my 500-mm lens, I

(Previous)
A majestic whitetail buck on a crisp, frosty morning.

(Above)
A large buck disguised among the branches of a thicket.

cautiously crept down the embankment into the frosty swale grass. By the time I had my tripod on the ground and my camera aimed at him, the buck had stopped rubbing. Gradually he emerged from the tangle of brush, and what I saw made me tense behind the lens. He was a magnificent animal, obviously in the prime of life. His antlers were so massive that he had to tip his head in order to maneuver through the branches in the thicket. I motored through my film, shooting as quickly as my camera and eye would allow. I had exposed four rolls of film when a second movement caught my attention. At the far end of the clearing two deer, a buck and a doe, burst from the forest. This buck was hot on the doe's trail, pursuing her at a brisk trot into the field. As they neared I could hear his grunts and make out his handsome eight-point rack; he was a nice, typical buck, but unfortunately for him, not the caliber of the bruiser still hidden in the thicket. The larger buck had, of course, also noticed this couple and their seemingly reckless, hormone-driven approach. The large buck took a moment to assess the situation and size up the other buck, then began to move from the thicket. As he stepped into the swale grass, the courting couple, now a mere thirty yards from him, came to a quick stop and stared at the new, more dominant suitor.

After seeing the big buck exiting the thicket, the wide eight-pointer starts to back away from the doe.

Ears back and hair bristling, the big buck approached his rival. Lowering his head and tipping it to the side as he neared, he made his point obvious. The big buck meant business. "Leave or prepare for the fight of your life" was the message. The eight-pointer showed little hesitation and backed off, slowly angling toward the forest edge. He knew he had a better chance of finding another doe in heat than he did of defeating this big fella in an all-out fight.

After the eight-pointer's departure, the big buck trotted toward the doe with head lowered, all the while making a deep guttural grunt. The doe, accepting her new suitor, exited the field and entered a nearby woodlot. The buck followed closely behind. With virtually all the leaves off the trees, I could follow them with my telephoto lens as they started up the forest ridge. It was only a matter of seconds before they disappeared.

Once they had vanished I stood up, stretched, and took a deep breath of the cool morning air. Everything was quiet and still again. I decided to follow the trail of the couple that had eloped into the woods. Upon reaching the top of the hill I found myself with an excellent view of a gully and ridge through the hardwoods. Since the rut was in full swing and this doe was obviously in heat, I decided to sit down and wait within sight of her

The large buck in the frosty swale grass looks my way before trotting after the doe and disappearing into the woods.

trail, hoping that other mature bucks would pick up her scent and follow it. If they did they would be close enough for me to photograph.

As I sat reliving the events that had just occurred in the lowland field, counting the potential photographs in my mind, time lapsed. The light was changing: the sky had been clear (a photographer's preference at sunrise), but now some high wispy clouds were moving in. This was great! On a clear day, as the sun gets higher in the sky the light will become too "harsh," creating stark shadows in the photographs. With an increase of light cloud—just enough to cover the sun, but with the halo of the sun still visible—the light will become even, and the shadows will disappear. This change in the light makes it possible to continue shooting through the late morning and early afternoon.

After I had been settled on the hardwood ridge for about half an hour, the buck came back within sight.

About thirty minutes passed before I heard movements approaching me. Leaves rustled and a couple of twigs snapped before I saw them. The same doe was heading back toward me along the ridge, followed by the big buck and, to my surprise, the persistent eight-pointer (at a respectful distance). I lowered my head to look through the camera, quickly set the shutter speed and aperture, and focused. The large buck had just stepped out from behind some trees, and I managed to compose and get off several pictures before the doe's scent lured him closer. He approached the doe, lip curling, testing the air for her sexual pheromones to determine if she was ready to mate. As he moved within a couple of feet of her she teasingly turned and trotted ahead of him, luring him further down the ridge and out of sight. Resigned to his subordinate position, the eight-pointer chose not to

Never wandering far from the doe, the big buck approaches her to see if she is ready to breed. The wide-racked eight-pointer satellites in the background, risking the wrath of the dominant buck.

follow, contenting himself by browsing. This second appearance lasted only a matter of minutes, but what a rush! Just imagine sitting in a tranquil stand of maple and beech trees enjoying the soft breeze, when three deer appear and charge toward you, oblivious of everything but their rutting instincts.

As it turned out, that would be the last of my shooting for that morning; they did not return again, and no other buck followed the doe's trail. As I looked around, appreciating the vastness and the intricate detail of the forest surrounding me, deep satisfaction in what had just taken place consumed me. Like all wildlife photographers, I work so hard to be "in the field," putting in so much research, investment, behavioral knowledge, and timing. All of these things, and of course luck, make these special photographic events profoundly satisfying. This is one of the key reasons why wildlife photographers are so emotionally driven.

The white-tailed deer is the most popular and abundant large mammal inhabiting the wilds of North America. Its sleek beauty, elusiveness, and complex behavioral patterns make it one of the most challenging and rewarding subjects to photograph.

Most people today are not lucky enough to see many of this continent's large mammals in the wild. Whitetails, however, inhabit most of the United States and southern Canada: wherever there is a tract of woods there are usually deer.

A deer is a glimpse of the genuine, exciting, untamed North America. People gasp when they spot one and, in that moment, feel close to something wild.

Every picture in this book is a record of a wonderful moment. Experiences in nature like these are spiritual to me. They make me feel whole. I hope that you feel something of the power of these encounters through the images and text in this book.

The big buck follows the fragrant doe as she wanders over another ridge. An unexpecting fawn looks on in awe as the large buck approaches.

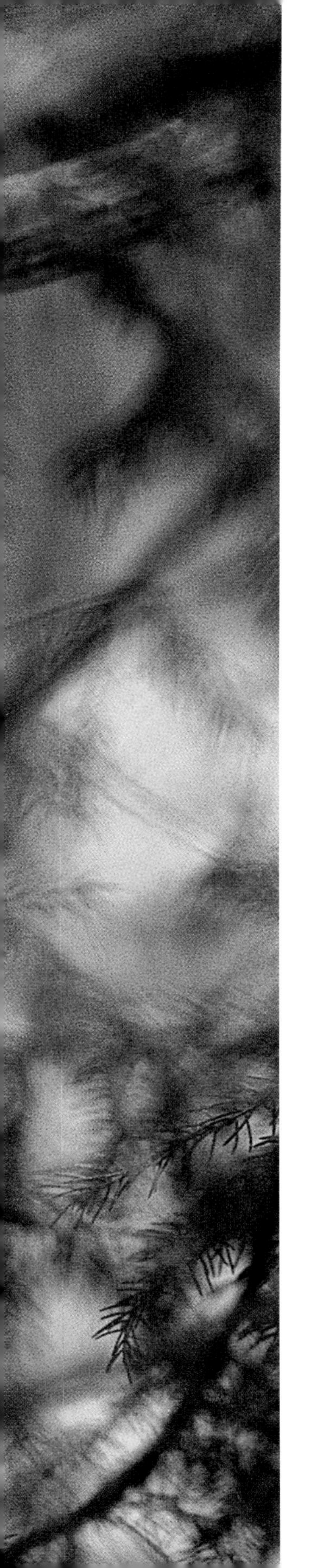

ONE

Evolution of the Whitetail

To us white-tailed deer are a symbol of beauty and wildness. To the rest of nature they are a rugged, adaptable survivor. Living and evolving in a diversity of habitats, feeding on a tremendous variety of plants, and developing incredibly acute senses in response to the most cunning predators have all led to the success of the whitetail.

Unlike many other large wild mammals in North America, whitetail populations have expanded over the past century. With the encroachment of humans into their furthest domain, whitetails are obliged to exist in close proximity to people—but surprisingly, their lives remain very secret.

Today, white-tailed deer are the most widely distributed mammal in North America. They can be found in all American states except Alaska and Hawaii, and from the Florida Keys north into Canada. They also roam as far south as Peru in South America. In Canada, their boundaries don't stop along the Great Lakes. Remarkably, white-tailed deer have been seen as far north as the Yukon and the Northwest Territories! Evidence of their expansion north of the sixtieth parallel is a testament to whitetail adaptability.

The ancestor of the modern whitetail first appeared in North America approximately four million years ago. Humans arrived approximately 11,000 years ago. Since then, deer have become an important part of the diet and folklore of North America's Native peoples. Used for food,

(Previous) The always-alert white-tailed deer has evolved to be one of the most successful large mammals in North America. There are more white-tailed deer today than ever before.

clothing, and ritual, the white-tailed deer has long been a part of human interaction with nature.

In North America there are three types of deer, all evolving from the same ancestor (*Odocoileus*, or "hollow tooth"). They are white-tailed deer, black-tailed deer, and mule deer. Each type of deer was named after a unique aspect of its appearance. Whitetails have a large, bushy, white tail (bearing brown and sometimes black hair on the outside), blacktails have a darker, black tail (but still have white hair on the underside), and mule deer have larger, mule-shaped ears (their tails are smaller and white with a black tip).

White-tailed deer are by far the most widespread and abundant of the three types. Blacktails and mules both prefer a western, mountain habitat, with blacktails selecting the densely forested mountain regions and mules preferring a more open, brushy, mountain terrain. Whitetail and mule deer do overlap and inhabit some of the same range in the Rocky Mountains, but they rarely mix. Whitetails prefer the forested river bottoms, while mule deer are most often found higher up in a more exposed habitat. One of the key reasons for these different habitats is that these two types of deer have evolved distinct styles of escaping predators. Whitetails run and run fast, leaping into thick forests, over deadfalls, and along well-known trails, often leaving the predator in the dust. Mule deer, on the other hand, escape in a peculiar-looking, springing "stott." Instead of flat-out running, mule deer bounce, bounding up steep ridges and over rocky obstacles impassable to the predator; the predator must go around these barriers and waste time.

On rare occasions, white-tailed deer and mule deer interbreed. When this happens, the offspring's chance for survival is slim. The hybrids don't seem to inherit the escape strategy of either parent. When confronted by a potential predator, they neither run fast, nor stott away; instead they act confused, giving their attacker ample time for a kill.

Of the three types of deer in North America, the whitetail is considered to have been around the longest, with mule deer being the most recent to evolve, about 8,000 years ago.

Although they are not found among the higher elevations of the rugged mountains of the northwest, the northernmost tundra, and some of the desert spaces in Utah and Nevada, whitetails occupy virtually every other habitat in North America. From deciduous forests to mixed pine/oak savannas, farmlands, swamps, and even bordering on deserts, whitetails have proven to be the most adaptable wild mammal in North America.

True to Darwin's theory of evolution by natural selection, whitetails may exhibit significant differences from one region to another. This drastic diversity of environments has caused deer to vary in body size, antler size, and coat color.

A whitetail's keen sense of smell, hearing, and sight are reasons why it has been so successful.

They have also adapted their behavioral cycles, altering their mating and fawning schedules to best suit their habitat and its specific seasonal changes. These marked differences between populations of whitetails have caused scientists to classify them into subspecies. The first such subspecies to be classified was the Virginia whitetail (*Odocoileus virginianus virginianus*), in 1780. This is likely where European settlers first described white-tailed deer (explaining the Latin name given to the species). Today there are as many as thirty-eight subspecies of white-tailed deer (thirty in North and Central America, and eight in South America). Mind you, this number is continually being debated among scientists. The argument is whether or not some of these subspecies are distinct enough to warrant separate classifications. There is, however, enough supporting evidence to warrant some splitting into subspecies, especially when one considers that in the Florida Keys a mature, breeding buck weighs about seventy-five pounds, whereas in Canada a buck of the same age and stature may weigh over 300 pounds. The large geographical separation of these two subspecies makes it easy to understand why they are so different. To cope with the cold winters of the north, the northern subspecies have become bigger; this common evolutionary adaptation is observed in mammals that inhabit the more northerly latitudes.

Also, the surface-to-volume ratio of deer in the north is less than that of southern deer. The larger, northern deer have a smaller surface area compared to body size, reducing the amount of heat loss in the colder climate. This evolutionary adaptation in mammals is referred to as Bergman's Rule.

Northern whitetails also tend to have slightly smaller ears than their southern counterparts. Smaller ears mean less heat loss during the cold winter months. Having smaller ears in the north may also reduce the risk of frostbite. Although very rare in deer, there have been isolated cases of frostbite. In fact, in one of my research areas in Ontario, I came across this oddity: a whitetail doe that had lost the upper half of both her ears from the cold. Despite their absence, she continued to lead a healthy, normal life for several more years, but I will admit she looked quite strange with only stubs on top of her head.

The subspecies that inhabit the warmer, southern climates benefit from having larger ears. Air flow over their thin, highly vascularized ears promotes bodily cooling.

In Canada there are three subspecies of white-tailed deer. In Ontario and eastern Canada there is the Northern Woodland Whitetail (*O. v. borealis*). This subspecies is the largest-bodied of all the white-tailed deer, weighing as much as 400 pounds. From Manitoba to British Columbia there is the Dakota Whitetail (*O. v. dacotensis*). In southern British

Whitetails have a magical way of standing still and blending in with their environment.

Columbia there is the Northwest Whitetail (*O. v. ochrourus*). As well as their distribution in Canada, these three subspecies cover most of the northern United States.

The whitetail still fascinates people today. It is the most researched wild mammal in North America, with scientists and biologists across the United States and Canada studying its complex behavioral patterns. They're striving to understand everything from antler development and uses to rutting biology, social interactions and hierarchies, and feeding and movement patterns. Whitetails are complex animals. Their acute senses and agility ensure that only the most astute and persistent predators may make a meal of them. Natural selection has shaped whitetails to perfectly fulfil their role in nature. Everything from their body shape to their four-chambered stomach perfectly suits their lifestyle as one of North America's most successful herbivores.

Structurally, white-tailed deer are built for speed. Their hind legs have very strong muscles that can launch them over obstacles eight feet high. Their legs are slender and strong, tapering to their compact hooves. For their body size, deer have small "feet" compared to other mammals. A whitetail's hooves are small and hard, reducing friction and allowing for faster getaways. The bottom, or sole, is the softest part, probably for purposes of added traction. The paired "toes" on each foot actually spread apart and spring back together with each step (helped by ligaments), adding to a deer's propulsion.

A deer's hooves are not the same as our toes. In whitetails, the lower half of each leg is simply an elongated hand or foot, with the hooves being the fingernails of the third and fourth fingers (our two middle fingers). Hooves are made of keratin, the same as human fingernails, and grow throughout life. Deer continually wear them down while travelling or running over the terrain. Above the hooves, on the back of each leg, are the dew claws (comparable to our index and "pinky" fingers). These appear in tracks when whitetails walk on soft ground, or when they run and put more weight on their hooves.

For centuries people have been trying to determine the sex and age of deer by looking at their tracks—a daunting task. Tracks left by male and female deer are very similar. Bucks on average weigh thirty-five percent more than does of the same age, and hence have larger hooves. The largest, most rounded-looking tracks belong to mature bucks. Younger bucks and does have a more defined, sharper edge and point to their tracks. Aside from the tracks of a very large male and those of a small fawn, it is very tricky to accurately pinpoint a whitetail's sex and age. The differences are very subtle.

Deer have survived for such a long period because of their ability to capitalize on available food resources. They have one of the most varied diets among herbivores. Food resources vary from region to region, but when considering the different plants available in a given locale, deer will consume most of them, selecting the optimum forage first. Being an edge species (hovering on the edge of two different habitats), deer feed on the available food in more than one habitat, most often forests and fields. This diversity in diet helps to ensure the survival of the species. If one of their favored foods is threatened, or disappears, deer will have many others to select from. As the land has evolved and some of the plants have changed over the millennia, whitetails have altered their diets, continually utilizing the highest-quality foods.

White-tailed deer belong to a group of plant-eating animals known as "ruminants." These mammals have a four-chambered stomach designed specifically to digest highly fibrous, vegetative matter. The first three chambers (the rumen, reticulum, and omasum) house micro-organisms (bacteria and protozoa) that break down the tough plant material called cellulose. In the first and second stomach compartments, ingested plant tissue is fermented by the microbes, reducing it to a softened mass. Like other ruminants (moose, domestic cattle, etc.), deer regurgitate the softened mass for more chewing, hence the term "chewing

Deer are very selective when choosing their bedding areas. This buck has bedded at a high vantage point with coniferous trees behind to shelter him. This location provides him an excellent view of the woods, to watch and listen for predators.

Deer are extremely adaptive. As humans increased the amount of farmed land across North America deer have recognized this new, high-quality food resource. In some areas agricultural foods are a key part of the whitetail's diet.

their cud." When the food is swallowed a second time it is sent to the third and fourth compartments. The third stomach cavity serves to pump the partially digested food to the fourth chamber (the abomasum). This final compartment resembles a "true" stomach, similar to that found in humans. Here, the strong acidity and the addition of enzymes aid in the digestion of the fermented matter, including any micro-organisms that have been passed from the first two stomach compartments.

Whitetails have evolved the teeth that best suit their diet. Deer have thirty-two teeth; most are premolars and molars, which are composed of hard and soft layers that wear unevenly, maintaining an edge for grinding plant matter. Deer also have lower incisors at the front of their mouths, but interestingly don't require upper incisors to nip off vegetation. In place of these teeth, deer have a bony pad. When foraging, deer will select the richest quality plants, bite them off and swallow, hardly chewing. Whitetails and other ruminants are prey species, and it is in their best interest to feed efficiently, minimizing exposure to predators. The chewing can wait until they're in a more secure bedding area.

Deer live by a simple code: eat, rest, reproduce, and avoid being eaten. Whitetails are "crepuscular" in their feeding activity. Unlike nocturnal or diurnal animals, which are most active at night or during the day, deer are most active at dawn and dusk. When not eating, deer spend a significant amount of their time at rest, conserving energy for when they'll really need it. Bedding locations are selected carefully: they provide cover, and often a view of the surrounding woods to watch for predators. In inclement weather, deer usually bed close to some form of shelter, most often conifer trees like cedar or hemlock. Whitetails tend to remain bedded through the mid-morning and mid-afternoon, with some feeding activity around mid-day. The majority of the time spent bedded is used to "chew their cud," interspersed with brief intervals of sleep, usually lasting a few minutes at a time. Deer commonly bed together in small groups, relying on each other to detect approaching danger while at rest. Come evening, deer once again become active, and move to their feeding areas.

Whitetails have managed to survive and adapt while other species have gone extinct. When deer first arrived in North America, the fauna present were considerably different than today. Deer didn't have to elude wolves, or humans toting spears, or later, guns. They were stalked by larger, giant carnivores like saber-toothed tigers and dire wolves that dwarf today's cunning canids.

It has been estimated that prior to the arrival of European settlers and their more modern weaponry, whitetails numbered more than twenty-five million. By the late 1800s, with the railroad and roads leading from the Atlantic coast west across the land, people gained increased access to the wilds of North America, and to whitetails. Venison became a marketable commodity and deer were harvested in excessive numbers. The meat was shipped via railroad back to the larger, developing cities to be sold. By the turn of the century it was estimated that less than 500,000 whitetails remained. With this dramatic decline, deer were becoming harder and harder to find, and fewer commercial hunters were continuing the harvest.

Early in the 1900s it had become apparent that some kind of control measure would have to be implemented to protect deer from disappearing altogether. Controlled hunting seasons and bag limits were put into place. Due to the success of these efforts, deer numbers have rebounded beyond expectation. Today, it is estimated that there are more whitetails in North America than there were when the first Europeans arrived.

The adaptable whitetail has benefitted from humankind's development of the land and agricultural practices. Whitetails have expanded northward in recent times due to the changes humans have caused in the environment. Humans have logged older, mature forests, creating younger

(Following) Why do deer have white tails? Once believed to serve as a warning device for other deer, this white "flag" is now thought to relay a message to the predator.

stands of timber, which have a denser understorey of preferred browse for deer. We have felled trees and planted fields of highly nutritious plants, like alfalfa or winter wheat. We have cleared extensive strips of northern boreal forest to erect pipelines, creating new edge habitats for deer. In many areas, especially in the south, we have removed most of their natural predators, such as wolves. With humans as their unwitting benefactors, is it any wonder that whitetail populations have fared so well?

The life span of white-tailed deer can vary tremendously from one region to the next. This regional variation depends largely on hunting pressure, whether or not there are other predators such as wolves, and seasonal fluctuations in weather patterns. In the wild, deer have been known to live as long as sixteen years. However, this is rare. Most deer live far shorter lives, succumbing to predation, disease, or environmental stresses such as harsh winters. Bucks lead a more active, confrontational lifestyle than does, resulting in a shorter lifespan; does usually live a few years longer than bucks. In captivity, with the absence of predators or food shortages, deer have lived for as long as twenty years.

White-tailed deer have obviously been named after the part of their anatomy that people see most often, their tail. When a deer is startled by a threatening sound, smell, or sighting, it will usually flee, all the while waving its large white "flag." Why do deer, and other herbivores for that matter, have white tails, or rumps? They were once believed to serve as a warning device to other deer; when a potential threat like a wolf was near, it was thought that a deer would raise its tail to alert others of the impending danger. Today, this white "flag" is understood to serve another purpose. The tail, which can exceed twelve inches in height, is now thought to relay a message to predators. The message it's sending is, "I have seen you, and I'm leaving; since I'm faster than you, don't bother following me." This forces the potential predator to make a decision right then and there: pursue the deer or give up and look elsewhere. Most predators will stop and move on, preferring to take their prey by surprise at close range rather than engage in an extended chase. To add further support to this theory, deer that are startled at close range don't raise their tail. Instead, they simply run away. Only when at a comfortable distance, feeling confident that it can outrun the threat, will a deer wave its "flag" and say goodbye.

Deer have incredibly acute senses. A deer's olfactory sense is probably its most important tool. Smell plays a very significant role in the many aspects of the day-to-day life of deer. Everything from detecting predators to finding and selecting foods or identifying other deer, from the bond between does and their fawns to

When deer are startled at close range they rarely lift their tail.

attractant pheromones (scents) produced during the mating season, are all assessed with a whitetail's nose. Deer regularly lick their noses; more scent molecules will stick to a wet nose than a dry one.

Throughout the year whitetails use their large noses to pick up pheromones produced from the skin glands of other deer. Pheromones are used to help deer communicate with one another. They may relay location, mood, or sexual readiness, warn of potential danger, or be used to mark territories.

White-tailed deer possess several glands which secrete pheromones. The most noted of these are the tarsal, interdigital, forehead, and preorbital. Located halfway up a deer's hind legs, the tarsal gland is argued to be the most significant scent-producing gland in whitetails. Deer urinate on these glands, creating a musky odor, likely communicating their dominance to other deer. It is also suggested that this scent helps deer to identify one another.

Interdigital glands are found on each foot, between the deer's "toes." They leave identifying scent along the deer's trail.

The forehead gland disperses scent when it is rubbed on overhanging branches. These branches serve as communication posts. Whitetails will also nibble or lick the branch, checking for scent left by other deer. Forehead glands are most active during the mating season, especially in males. They are also used by bucks to deposit scent on trees that they rub prior to and during the breeding season.

The preorbital glands, found in front of the animal's eyes, are also used for scent dispersal. The exact message that this chemical cue is relaying remains unknown.

Hearing is the second most important sense used by deer to avoid predation and to communicate with each other. Their large, cup-shaped ears provide them with long-range hearing. In fact, deer can rotate their ears independently and point them in different directions without moving their head. Since deer are a prey species, it is important to their survival that they can hear as well or better than their predators. Whenever hearing is impaired, sometimes due to high winds, deer will remain bedded, minimizing movement.

Whitetails have good vision; they can pick out movement at amazing distances. The location of their eyes on their skull allows them to see 310 degrees without moving their head. People, on the other hand, with eyes centered on the front of the skull, have a considerably narrower field of view. This ability of whitetails to monitor virtually all of their surroundings without even moving, picking up the smallest unnatural movement, helps them minimize being detected by predators.

Deer can also see at night. Their eyes have an extra membrane which reflects light back through their retina a second time. This membrane is called a "tapetum lucidum." Many people have seen it: whenever deer are caught in the headlights of an oncoming vehicle, the reflection or shine that is seen is this additional membrane. All mammals that are active at night have this reflective tissue, including your pet dog or cat.

The most debated subject about a deer's visual sense is its ability to see colors. The mammalian eye has two types of receptors, rods and cones. Rods enable animals to see in low light, whereas cones perceive color and bright light. A deer's eye is thought to have about half the number of cones of a human's, suggesting a limited perception of color. Deer are not truly color-blind but rather see a narrower spectrum of colors than humans. For instance, they tend to see reds and oranges as similar to earth tones like brown.

Along with their well-honed senses, whitetails have developed an elaborate system of communication. Aside from the odorous chemical signals discussed earlier, deer use vocalizations and body language to relay messages.

Whitetails emit an array of vocalizations. These include various types of grunts, snort–wheezes, snorts, bleats, and bawls. This list may not seem extensive, but different ages, sexes, and sizes of deer cause these sounds to vary in pitch and tone, creating a broad assortment of calls.

The most common sound emitted by bucks is a short, guttural grunt. The more mature the buck, the deeper the grunt. Bucks often grunt when approaching other deer, making their dominant presence known. They are most vocal during the mating season, while pursuing or tending a doe. While approaching a doe the buck will emit several short grunts followed by a louder extended concluding grunt. While emitting this last grunt the buck will usually run at the doe, chasing her a few yards. A dominant male in the company of a hot doe may also make an elongated, soft, tending grunt.

When extremely agitated, bucks will make a snort–wheeze vocalization. This sound is usually emitted by mature, dominant bucks prior to attacking another male. The buck will inhale and exhale quickly several times, making a loud breathing sound, followed by one drawn-out inhale. This is one vocalization that people can easily reproduce. If you clench your teeth, frown (this draws your face tighter), and breathe in and out quickly several times (through your mouth), followed by an extended inhale while you straighten up your body, you will sound like a very angry whitetail buck. I recommend that when you try making this sound, you do it out of sight and earshot of other people, just so they don't think that stress has finally gotten the better of you!

The most common deer vocalization heard by people is the loud alarm call, or snort. This sound is made when deer force air through their nasal

White-tailed deer have evolved high rates of reproduction. Does can produce one, two, or, on occasion, three fawns per year. This high rate of fecundity enables deer to cope with relatively high rates of mortality due to predation and, in the north, harsh winters.

passages. I'm sure that more than one person has been startled by the sudden snort of a hidden deer that has smelt or heard them. Not unlike the sudden explosion of a ruffed grouse taking flight from underfoot, the sound of a whitetail's alarm call seems to stop time for a second or two.

Deer snort when they sense danger. Similar in function to their white tail, the call serves to alert potential predators that they've been found out. Snorts may also serve to notify other deer of an impending threat. Does seem to give this bursting alarm more often than bucks.

Bucks and does live most of the year in separate social groups. Doe groups are made up of related individuals: a matriarch doe, her daughter or daughters, and granddaughters. Since they are genetically related, it is in their best interest to alert one another of danger and protect their genetic investments. Buck groups, on the other hand, are composed of males that are almost always unrelated. With the blood-bond absent, a buck will tend to exit a threatening situation with minimal detection, saving himself with no worry spent on the other males.

From my experience, deer don't snort until at least two of their three primary senses have been triggered. In fact, it is common for deer to become quite curious and approach you if you have only triggered one sense. In other words, if a deer has heard you but can't see or smell you, it may move closer, trying to figure out whether you're friend or foe. In fact, this is how I met my wife. One evening while I was out in a field photographing a young buck, my future wife was cycling past on a road about one hundred yards away. Looking up, she saw a man standing and focusing a large telephoto lens, and a young buck slowly stepping toward him. She watched while the buck approached to within thirty yards before turning and prancing away. She waited for me to exit the field; she couldn't resist asking why that deer had approached me. Was I some kind of deer whisperer? I explained that I had been lucky enough to trigger only one of the deer's senses. Like most young animals, it was curious, and decided to come nearer and investigate me. We talked for a while, and the rest is history.

Another sound made by deer is a bleat. Does and fawns make bleats, with those of fawns being higher pitched. Does may bleat when beckoning their fawns to emerge from their secret beds and nurse. Does also bleat at other deer when expressing their dominance. This is usually over a food source. A mature doe will utter a bleat to suggest to a subordinate doe or yearling buck to move aside and allow her to browse. Fawns bleat most often when trying to get the attention of their mothers, usually when they're hungry. Fawns don't emit calls during their vulnerable first few weeks of life. They spend most of this time silently bedded, hiding from predators.

Another vocalization made by deer is a bawl. This is a rare cry. It is loud, not unlike a sheep's "baaa," and is only emitted when a deer has been seriously injured.

Whitetails also communicate with gestures. One in particular is "stomping." When deer are suspicious of a nearby presence, such as a person walking by, they may stomp on the ground with one of their front hooves. When this happens the deer has probably had only one of its primary senses triggered, and remains uncertain of what may be nearby. The stomping may serve to encourage the potentially threatening animal to show more of itself, allowing the deer to better assess the situation. The message deer are relaying is, "I know you're out there—what are you? If you're a predator, give up, because I'm on to you." Once they sense a possible danger, deer focus on it until they know exactly what it is. Mistakes in their world are too costly!

Other gestures exhibited by deer include displays of dominance. These include: ears flattened back against the skull; hair bristling on the neck and back; a slightly sideways, aggressive walk toward a subordinate; and, eventually, front hooves lashing out or antlers plowing into the challenger. Dominance displays usually occur over preferred food resources, or in male–male confrontations, during the mating season.

Northern whitetails have adapted to the most variable environment inhabited by deer, an environment where at least three months of the year are consumed with cold temperatures and icy snow. Aside from having a larger body, these deer also have hollow hairs. Every hair in their coat has a small pocket of air, providing superb insulation. Their winter coat is so effective at retaining heat that when they are at rest, freshly fallen snow may lie on their backs for hours without melting.

Whitetails have also adapted behaviorally to cope with winter. When snow levels reach close to a foot deep, deer will migrate to a small wintering range. These wintering areas, which house concentrations of deer, are commonly referred to as "yards." Whitetails are drawn to yarding areas because of their available shelter and winter food. Consisting of stands of evergreens like cedar or pine, yarding sites shelter deer from some of winter's wind and drifts. The needled bows of these trees act as a wind break catching the cold, blowing snow. When several deer bed in close proximity to one another under these evergreens, the bows will trap some of the heat escaping from their bodies. In ideal situations this heat-trapping effect creates a warmer micro-environment for the deer.

Whitetails spend most of the winter bedded, restricting their movement and conserving energy. In fact, a whitetail's metabolism slows significantly during these harsh months, decreasing the amount

Deer have expanded their range northward, living in areas as far north as Alberta. Apparently, white-tailed deer have even been spotted in the Yukon.

Deer in the north have changed their seasonal behaviors to survive the limited food resources and the frigid temperatures of the winter season.

One of the primary adaptations leading to the success of deer in the north is "yarding." When the land freezes and the snow is deep, deer will group together in parts of the forest thick with evergreen trees. These trees provide shelter and the deer create worn paths through the snow for easier travel.

of food required and allowing the deer to remain still, reducing exposure to the elements. When bedded, deer tuck in their hind legs and often curl up in a ball, similar to the way a dog will. This exposes a minimal amount of surface area to the cold. When winter brings a day or more of continual snowfall, deer will remain bedded, waiting for the inclement weather to pass before feeding.

Throughout winter, forage is limited. As snow covers the forest floor and autumn foods like apples disappear, whitetails must switch to woody browse. Far less nutritious, winter foods like cedar, hemlock, and deciduous twigs make up most of their diet. Combined with this lower-grade winter diet is increased competition. With many deer sharing the resources found close to the yard, food supply may run low. If winter persists and foods are depleted, younger, subordinate deer, like fawns, may die of starvation.

To cope with deep snow, whitetails make trails around and through their yarding area. Packing down the snow makes for easier travel and faster

The white-tailed deer has adapted to an ever-changing world and is the most successful and prolific large mammal inhabiting the wilds of North America.

getaways from predators like wolves. The trails, sometimes referred to as "runways," get packed from repeated use as deer move among the yard while feeding. In the absence of trails, the pointed hooves of deer plunge deep into the snow and they become much more cumbersome and vulnerable. Most northern predators have large feet padded with tufts of hair, preventing them from sinking as deeply into the snow. When a thin layer of ice forms on top of the snow predators have the advantage. A whitetail's hooves will break through the thin ice, where predators will often run on top. The maintenance of runways is critical to deer survival in deep snow.

Early one January I was photographing an impressive ten-point buck as he maneuvered through a wooded area along a packed trail. Once he was out of sight I decided to follow him on the runway to see if I could get some more shots. After rounding some evergreens I saw him about fifty yards off, but he seemed to be on a different trail. The only way I was going to get close enough was to walk across the snow to the next trail and follow it toward him. Well, I had no idea how packed the snow could be on a deer runway. I stepped off of the trail and almost disappeared—the snow was up to my chest! Needless to say, I couldn't pursue the buck. It was hard enough just to climb back up onto the trail while toting pounds of camera gear. Deer runways do facilitate winter travel.

Today, whitetails continue to evolve. Their predators, food resources, and habitats have altered over the past four million years, and they themselves have adapted. Entrancing, elusive, and complex, whitetails continue to fascinate humans. The fawns with their large eyes and dappled backs have captured the affection and imagination of children and grown-ups alike. The graceful doe, lithe and lovely, often trailed by her fawns, shows us that mothering is a critical element of species survival. In the whitetail buck, we find the essence of wildness made hide and sinew, and a crown of bone uniquely formed on each and every bearer. These crowns unfailingly evoke curiosity and wonder.

TWO

The Antler Cycle

As the sun dipped toward the horizon, a group of does were browsing in a field with their fawns, enjoying the warmth of an early September evening. Their ears were tuned to the slightest hint of danger. As they fed in relative safety, they heard something moving through the forest toward them, and their bodies came to full alert. However, their acute senses quickly relayed that the source of the noise was not a threat, but instead another deer. They stared intently in the direction of the sound until a large buck stepped from the shadows.

The buck noticed the cluster of females but did not approach them; instead, he headed across the clearing in the direction of some shrubbery. His rack was fully grown, the velvety tissue once used to nourish the developing bone now dried and wrapped tightly to the antlers. As the buck reached the shrubs, he lowered his head and ground his antlers against the bark. The velvet cracked and tore from the bone. Over the past few weeks the buck's restlessness had risen considerably and he was eager to tear off the expired tissue and reveal his new rack. It seemed as though he was venting a summer's worth of frustration on the unsuspecting thicket, its branches bending under his weight.

Pausing, the buck lifted his head to confirm the safety of his surroundings. He looked past the strip of velvet hanging from his antler and saw that the does were still feeding peacefully. Lowering his head again, he continued rubbing. As the shreds of tissue fell from his rack, the buck ate them, with the exception of a couple of strands left clinging to the branches. With

(Previous) A mature whitetail buck, proudly displaying his fully grown rack of antlers.

the bush swaying at the buck's mercy, the velvet was removed in about an hour, leaving his blood-red antlers exposed for the first time. Turning toward the does, the buck raised his head to the dying sun. Once again he was the monarch of the forest, ready to face any challengers and prepare for the upcoming rut.

Antlers, one of the most powerful symbols of wildness, have graced the North American forests since the retreat of the last ice age. Primitive people used antlers in religious practices as well as for tools and weaponry. In modern society antlers are used worldwide in a variety of forms, ranging from handmade carvings to pharmaceuticals. Aside from their tremendous appeal to humans, these elaborate crowns serve a more important role in the reproduction and survival of the animals that wear them.

Antlers are grown and shed annually from the heads of male white-tailed deer, or bucks. This cycle begins each spring, when budding antlers emerge from a bony pedicel located on each side of the skull. Whitetail bucks that live in the northern part of their range start growing their antlers in late April or early May. In the southern United States antler growth may commence as late as June. More mature bucks start growing their antlers first, followed shortly after by younger males.

In the early stages antlers appear as short, bulbous antennae, but it doesn't take long for them to divide into the main beam and tines. Antlers are the fastest growing bone in the world, with adult males growing up to half an inch of new tissue per day! This extraordinary growth rate is sustained by the outer suede-like component of a growing antler, called "velvet." This sensitive tissue ensheathes the rack and supplies the extensive blood circulation necessary to nourish the developing bone.

Antler velvet has another interesting feature: it is the only mammalian skin to regenerate hair follicles after birth. This soft tissue, which covers the antlers of the entire deer family, can somehow annually produce countless short hairs on its surface. In all other mammals, humans included, hair follicles cannot be naturally regenerated once lost through injury. In a sense, the shedding of a whitetail's rack each year can be seen as an injury, because when the antler is dropped the exposed pedicel must heal before the next antler can form. In humans, whenever an injury occurs to the skin, it heals over with scar tissue that doesn't contain hair follicles. Scientists have conducted studies to investigate whether this regeneration of hair follicles in velvet can be applied to human hair loss. This would certainly interest many middle-aged men with receding hairlines. Unfortunately, due to the different physiological systems of deer and humans, it isn't possible at this time.

Antlers should not be confused with horns.

A yearling buck, his spike-like antlers ensheathed in nourishing velvet.

The suede-like velvety tissue that nourishes a buck's developing rack is covered with tiny hairs.

Horns are never shed but instead grow continuously throughout life from the heads of animals such as bison and bighorn sheep. Horns are made of a complex protein called keratin and have a bony core, whereas racks of antlers are made solely of bone. Horns may also be found on both male and female animals, while antlers are found only on males—with the exception of caribou. Both sexes of caribou grow antlers, but the cow's rack is considerably smaller than the bull's. However, there have been a few very rare instances where whitetail does have developed small, spike-like antlers. This uncommon phenomenon is believed to be caused by hormonal imbalances.

Antler development depends on three factors: heredity, age, and nutrition. Heredity is primarily responsible for the shape of antlers. Each buck develops a unique rack. Like fingerprints, no two are identical—from a thin-racked six-pointer to the magnificence of a giant "non-typical." The best way to identify individual bucks is to study antler configuration. Uniformity, width, height, number of tines, thickness, and length all enter into the equation. The inside spread between the antlers of some adult bucks can exceed twenty-five inches! Other males may grow tall, vase-like racks that seem to reach for the sky. The shape of a buck's antlers normally remains the same throughout its adult years, varying only slightly in size and in number of tines.

The size of a whitetail's rack is largely determined by age. During their first year, male fawns, or buck fawns, grow small bony lumps which are referred to as nubbins, or buttons. These protuberances usually remain covered by the thick mat of hair on the fawn's forehead. By their second autumn, one-and-a-half-year-old, or "yearling," bucks produce their first set of visible antlers. These racks are normally quite small, ranging from two individual points (spikes) to as many as nine points (total number of tines, or points on both antlers). The majority of these adolescent whitetails develop two- to four-point racks. The few that develop eight or nine points have thin, tight racks that are small versions of those carried by adult bucks.

A "point" is any part of the antler that is taller than it is wide, and at least one inch long. The number of points on a buck's rack has nothing to do with age; rather, it is the *mass* of the antlers that changes as a buck matures. The size of a buck's rack increases significantly each year until he is about five and a half years old. Bucks that grow the largest, heaviest antlers are those that are in the prime of life, between five and a half and eight and a half years of age. These dominant males develop wide, sweeping, thick-beamed racks because they have reached their optimum body weight and are able to devote large amounts of their summer energy to the expanding antler bone. Younger

The following two photographs show how a maturing buck's antlers can change from one year to the next. Pictured on this page and the next is the same buck at two and three years of age.

bucks must invest a considerable amount of summer energy into their growing bodies, leaving less for antler development. Older males that have passed their physical peak must now cope with the energetic demands of bodily deterioration, whether due to wear and tear or disease. These wise, old, gray-faced warriors tend to develop smaller, less uniform racks than they did during their prime.

The abundance and nutritional quality of available food resources also affects antler size. Bucks that inhabit areas with less than optimum-quality food resources, or areas that have been overbrowsed due to high deer populations, will develop smaller racks than males in areas that support high-quality resources. Agricultural areas containing crops such as clover and alfalfa are prime examples of a high-quality food resource. Young forests also provide whitetails with a fine selection of nutrient-rich browse. Regenerating woodlands that have a thin upper canopy—possibly due to selected logging—will allow enough sunlight to pass to the forest floor to

encourage a high density of plants and saplings. Mature, undisturbed forests with dense canopies inhibit the penetration of light and have fewer broad-leafed plants on the forest floor. In the north, these mature forests are more likely to contain moose than deer. In summer, moose commonly feed on aquatic plants in ponds or marshes that border mature forests, and are not as dependent on woodland resources.

When comparing the antler size of bucks that live in agricultural areas to those in extensive, unbroken tracts of forest, it is easy to see how important nutrition is to antler development. This is not to say that bucks living in large, continuous pieces of woodland grow dwarfed antlers; they can and do grow large, majestic antlers when in their prime. However, bucks that live in smaller tracts of forest, bordering farmed fields, have a tendency to grow heavier racks.

While in velvet, whitetail bucks are continually tormented by bloodthirsty insects such as deer flies and mosquitos. Can you imagine having

A mature whitetail buck demonstrates secretiveness as he drinks from the edge of a summer stream.

such sensitive tissue perpetually exposed to the bite of voracious insects? This may be one reason why reclusive male deer can sometimes be viewed in open clearings during the summer. Open areas that contain few trees have a faster air flow than dense forests, reducing the number of biting flies. Bucks will sometimes bed down in grassy clearings where there are fewer flies, re-entering the forest only when forced by the hot summer sun or the scent of a predator. On warm summer days I have watched bucks try to cope with a swarm of black flies or several deer flies buzzing round and round their head. The beleaguered animals continually swipe their antlers with their large ears, trying to brush away the flies. I have seen bucks break into a run while shaking their head to rid themselves of flies. When the sun sets and the cool night air flows through the meadows and forests, bucks find relief from all the biting flies that prefer the warmer temperatures of day.

Throughout the summer, while the male hormone testosterone is still at low levels, bucks live a very shy and passive existence, carefully protecting their tender antlers from injury. Growing antlers contain thousands of tiny nerves that make them very sensitive. Although rare, injury may result from a direct blow to the growing tissue, causing it to become deformed. If severe, this injury may be "remembered" from one year to the next: a buck may grow the same deformed antler for two or more years following the accident. Such injuries are not common, however, because bucks somehow know the dimensions of their antlers, enabling them to maneuver through trees without harming their bony growth.

Antlers may also display a phenomenon known as contralateral growth. When an area of a buck's body is injured, the antler on the opposite side may become deformed. Deformed antlers are often referred to as "non-typical" because they lack the "typical" symmetry normally observed on healthy bucks. A deformed antler may be shaped like an uneven ball with a few small spikes protruding from it, or it may be directed down or up, instead of curving out in the normal fashion. An antler that has been injured and deformed is almost always smaller than the normally developed antler on the other side of the buck's head.

"Typical" racks usually have the same number of tines, or points, on each antler. Racks with a total of eight or ten points are the most common for adult bucks. The tine length, or height, is usually quite symmetrical on "typical" racks. Despite a strong similarity between the two antlers on a "typical" buck, every antler is unique. That's right, no two antlers are the same, not even the two antlers carried on a buck's head in the same year. The tines may be longer, or thicker; there may be more tines; the tines may be branched; the main beams may be heavier or longer. In essence, there are countless variations when it comes to antler size and shape. However, most bucks carry "typical" symmetrically shaped antlers.

"Non-typical" antlers may sometimes develop on healthy males. This very rare phenomenon is believed to be controlled by heredity and is most often observed on older bucks that have passed their physical prime. Once a healthy buck starts to produce a "non-typical" rack, he usually continues to do so for the rest of his life; the rack becomes more and more unusual each year.

Different tines on a rack, such as an extra brow-tine, or a drop-tine (a tine that points down towards the ground), or a tine that is branched, are usually not enough to warrant a "non-typical" designation. "Non-typical" racks often develop considerably more points than do "typical" racks and have a disproportionate appearance. A twenty-pointer would be glaringly conspicuous among other bucks, and would be quite a sight! At its most extreme, a "non-typical" rack can resemble a bristling chunk of driftwood. Very few people are lucky enough to see a rare "non-typical" buck.

I remember when I saw my first "non-typical" (I have seen only a couple during my travels). I was driving down a single-lane forest access road in eastern Ontario. It was dusk and I was checking for deer in the fields when I drove past some tall shrubbery, looked to my right, and almost went into shock! There, a mere ten yards off the road, was the biggest buck I had ever seen. I was so stunned by the sight that I just kept on driving! About thirty yards up the road I stopped the truck, sat for a couple of seconds, and tried to decide how I was going to get back to him. If I got out and walked slowly up the road he would be sure to hear me and disappear. If I backed the truck up the same thing would likely happen. I chose the latter. Craning my neck as the truck backed up to the shrub, I could see him—he had moved another ten yards from the road. I put the truck in park and turned off the ignition. Placing my sweater on the windowsill, I balanced my telephoto camera and started to focus on the buck. Meanwhile, he had not taken his eyes off me. He fled and disappeared in a blink of an eye, escaping into a pine plantation, before I had a chance to take the picture. But in my mind the picture will always be vivid.

By mid-August the antlers of northern bucks will have reached their maximum size. As the days begin to shorten, and the summer nights start to grow cooler, the testosterone level of the bucks begins to increase. This hormonal charge causes the vessels supplying blood to the antlers to dry up, resulting in the solidification of the bone.

By September the velvet of northern whitetails has dried and "shrinkwrapped" to the fully hardened antlers, and is ready to be removed. During the first week or two most of the mature bucks will strip the velvet from their racks. Younger bucks start feeling their jolts of testosterone slightly later than their adult role models and usually don't start shedding until the middle of the month. By the third week of September, as the first hints of autumn color paint the forests, virtually all healthy bucks will have removed the spent tissue from their racks. The velvet is sloughed off when bucks rub their antlers on trees and shrubs, a process that normally takes from one to twenty-four hours. The velvet is often eaten by the owner, but sometimes ribbons can be found dangling from a branch.

A newly exposed rack will have a reddish color for a few days following velvet shedding. The amount of bleeding that occurs during velvet removal influences the color of the antlers. Usually there is not much bleeding because all of the blood vessels have dried up, leaving only the velvet sheath. The bark from the tree used to remove the velvet may also stain the antlers.

The time of velvet shedding marks a tremendous change in buck behavior. By now, testosterone

levels will have begun to skyrocket, and instead of being shy and passive, the bucks become increasingly bold and restless in anticipation of the upcoming mating season, or rut.

Bucks are now able to put their antlers to use. The first priority is to establish a dominance hierarchy. This critical mechanism of the rut enables bucks to rank themselves in order of fitness. In the north, this is accomplished from mid-September to mid-October, when bucks follow an instinctive code of male–male confrontation. They engage in controlled sparring, gently placing their antlers together and pushing until one buck is found to be stronger. Two bucks may spar several times, with matches usually lasting a few minutes or less. These bouts test a whitetail's physical condition and rank him in the dominance hierarchy. Bucks establish this ranking prior to the eruption of the rut and are thus able to minimize the number of life-threatening fights that can occur over a doe in heat.

Most pre-rut sparring matches occur between bucks that have spent the summer months together in social clusters called bachelor groups. These all-male fraternities can vary in size from

Throughout the summer bucks form social clusters called bachelor groups. These groups may vary in size from two to several males.

When the velvet is stripped from a buck's antlers, some bleeding will often occur. This influences the color of the rack. Since the antler bone is fully hardened and the velvety tissue is dead, the buck likely feels no pain when shedding.

two to several members of approximately the same age. Mature bucks prefer to keep company with individuals of their own stature, leaving the younger males to form their own small units. Occasionally an estranged younger buck, such as a yearling recently chased off by his mother, will innocently approach a mature bachelor group. The older bucks typically don't accept the attentions of these youngsters, and chase them off. However, during the pre-rut in early fall, adult bucks do become a little more tolerant of these untried males and will even spar with them. It seems to be quite a gesture when a mature, heavy-racked buck lowers his head to spar with a thin-racked yearling. The big buck will barely push and usually only needs to use one side of his antlers to fend off the yearling. From my observations, despite these occasional severely imbalanced matches, bucks most often spar with other bucks of approximately the same size.

As the rut approaches, the intensity of the sparring increases, and the dominant bucks develop a low tolerance for other males. At this point the bachelor groups break up for the duration of the mating season.

Aside from their instrumental use in establishing a dominance hierarchy, antlers are also used to create rubs on trees and shrubs. Rubs are

This buck was photographed at sunrise on September 5. He had obviously spent part of the night rubbing his dried velvet from his hardened antlers.

Once the velvet is shed bucks will spar, using their hardened antlers to test each other and establish a dominance hierarchy. Most sparring occurs between members of the same bachelor group.

made during the mating season, when bucks place the rough base of their antlers against a tree and move their head up and down, removing the bark. This signpost serves to alert receptive females, and warn other males of the buck's presence.

Antlers serve a third and very important role—to attract does. Antlers clearly indicate the health and vigor of their bearer. When a buck approaches a receptive doe it is believed that she will visually assess his fitness, which is exhibited by his body and antler size. Females will choose to breed with males that have the largest body and antlers. Only the fittest males can grow the awe-inspiring racks that instill dominance and guarantee reproductive success.

For decades it was believed that bucks that produced spike antlers as yearlings would never grow a large rack later in life. This is a misconception; young bucks that have only small spikes at one and a half years of age may still grow large, dominance-instilling racks as prime-aged adults. As yearlings, these bucks may have been limited to producing spike antlers by environmental stresses. Food resources may have been poor

At last, after several months of growing, this buck's antlers are velvet-free and ready to serve him in the upcoming rut.

When the rut has concluded and, in the north, winter grips the land, a buck's majestic crown is no longer of use. Mature bucks like this one, who breed most of the does during the mating season, are the first to drop their antlers.

during the antler-growing season due to drought, or the best food resources may have been over-browsed due to high deer densities. Any stresses experienced during the growing years of a young buck take their toll on antler growth. Body growth takes priority among bucks, especially those who are not yet fully grown. However, it is known that yearling bucks that produce two or more points per antler are more likely to produce a large rack at maturity than spike bucks.

Despite common belief, antlers are rarely used as weapons against predators. The principal role of antlers is in male–male competition for receptive females. They are reproductive status symbols. The fact that females do not grow antlers, and that males shed them after the mating season, support this theory. If they were important for predator defense, both sexes would carry them, and it would be a tremendous disadvantage to cast them aside each winter. Instead of using antlers, these large herbivores use their hooves for defense when cornered. However, the best defensive strategy of whitetails is simply to run, and run fast! Incredibly agile, deer can run thirty-five miles per hour and disappear through dense cedar swamps almost as quickly as they can cross an open field. This is the main reason why predators such as wolves fail most of the time when they attempt to capture deer.

As the northern woods become white with snow and temperatures plunge, the testosterone levels of the bucks drop dramatically. As the days once again become longer, shedding more sunlight into the northern woodlands, the connections between the antlers and the skull weaken, and eventually the rack is shed. The most fit, dominant males drop their antlers first. These monarchs were the most successful during the breeding season, causing their testosterone levels to drop sooner than those of younger or older, less fit males. Lower-ranking bucks may retain their antlers as late as February. Both antlers are usually cast within a day of each other. By dropping their antlers bucks relieve themselves of the extra weight, conserving energy during the harsh winter months. A large set of deer antlers can weigh as much as ten pounds! The purpose of this annual cycle is to enable bucks to grow a fresh set of antlers that best represents their fitness level for that year. This is especially important to young bucks, who eagerly want to climb the social ladder and impress the does.

Once antlers fall to the ground they are of no use to deer. They do, however, become useful to many species of rodents, such as porcupines, squirrels, chipmunks, and mice. These rodents gnaw on the fallen antlers, taking advantage of the rich source of calcium. Depending on the rodent densities of an area, deer antlers may remain on the forest floor from one to several years.

One day in early March, after the snow had melted, my wife and I were out exploring an area

This buck has shed one of his antlers, exposing the bony pedicel where the antler originated. The other side is sure to fall soon, probably within a few hours or a day.

of the woods frequented by several bucks during the winter. We were searching for antler sheds. Although I had searched in vain for about four hours the day before, my wife walked into the woods and found a shed in less than five minutes! The surprise was that the shed was from the year before, and I had walked within twenty yards of it on several occasions without looking directly at it. This area housed few rodents and the antler was practically perfect, with the exception of some green coloration. In another instance, late one autumn, I found the remains of a handsome antler. It had been shed directly beside a red squirrel's home in a pine plantation. Within the year the little squirrel had chewed away virtually all the antler except for the thick base. In this way one of nature's most extravagant gestures, the elaborate antlers of white-tailed deer, are used up with perfect economy in the cycle of life.

Once dropped, a buck's antlers are no longer of use to him. However, nothing in nature is wasted. Rodents such as this chipmunk will slowly gnaw at the antler, benefiting from the rich source of calcium.

THREE

The Rut

In the pre-dawn light of a frosty October morning, the large buck rose from his bed. He had been bedded for several hours, chewing his cud, before a long-awaited scent reached his nose. It had been many months since he had smelt the alluring fragrance of a doe in heat. The last time such odors filled the air he was not the dominant buck for the area, but over the summer he had grown considerably and was now in his prime. His rack was wide and heavy; he was a true monarch.

As he stood in the forest, his body charged with adrenalin, only one thing entered his mind: where was the doe? He had to get to her as soon as possible to prevent any of the lesser bucks from breeding her. He tipped his head back, lifted his upper lip and inhaled deeply, testing the air to discover which direction the scent had been blown from. Proceeding in a quick trot, head low to the ground, he travelled up a nearby hardwood ridge and intersected the doe's trail. Following the trail brought him to a grove of cedars. Grunting his presence as he tipped his head sideways, he maneuvered through the thick evergreens and stayed on her trail. When he emerged at the far edge of the grove he saw the doe; she was slowly exiting the woods at a nearby field. The big buck followed her to the edge of the forest and stopped. He stood dead still, watching the doe as she entered the field, and as he listened to the silence of the forest, he checked for any strange sounds or smells. After a few minutes, when no threats were revealed, the buck stepped from the forest and into the field, the dawn sunlight accentuating his heavy rack. He paused only

(Previous) Each autumn mature whitetail bucks focus all of their energy on courting and mating with does.

(Above) A mature, ten-point buck investigates a field of does, hoping that one will be in heat and soon be ready to mate.

briefly, then trotting quickly, he approached the doe. Remarkably, he was the first contender to find her. Upon recognizing his obvious health and stature, the doe accepted her suitor. The rut had begun.

Each year, shortly after the forests explode in color and the distant call of migrating geese echoes across the fields, white-tailed deer enter an annual state of frenzy called the rut. In the north this peak period of breeding activity occurs from mid-October to mid-December, with the majority of the breeding concentrated in mid-November. Whitetails found toward the southern portion of their range breed later than their northern counterparts. Without the pressure and concerns of an oncoming winter, deer in states like Texas typically breed from mid-December to mid-January, and sometimes even later.

During this time, countless people encounter deer while hiking, hunting, or driving to work. A glimpse of these elusive creatures can be awe-inspiring—unless, of course, in their hormonal intoxication they have collided with the hood of your car. Knowing that the white-tailed deer is

Each fall the food resources for deer change. The lush herbs of summer have succumbed to frost. Now, fruits like apples and forest nuts like acorns and beech nuts attract deer. This buck is feeding on freshly fallen beech nuts.

This buck is feeding on small white fungi, or mushrooms. Deer will frequently consume edible forest mushrooms.

normally a cautious animal, the average person may be surprised by these sudden displays of recklessness. This annual hyperactivity serves to ensure that only the fittest animals pass on their genes to the next generation. The pressure to breed is immense.

This time of year marks several significant changes in the lives of white-tailed deer. Aside from the mating rituals, deer in the northern half of the U.S. and Canada must also adapt to a changing climate and changing food resources. The succulent broad-leafed plants of summer dry and harden as frost-laden nights end their year. New sources of food must be found. Through late summer and fall it is critical that whitetails "bulk up," gaining weight to endure the harsh winter months ahead. Fall foods provide deer with the fat-building calories essential for this weight gain. Fall fruits such as apples and forest nuts like acorns and beech nuts are magnets for deer during the autumn months. Agricultural foods still standing in the fields through the autumn months also attract deer. Crops like corn and alfalfa are at the top of the whitetail menu. Various forest fungi, or mushrooms, are also consumed by deer.

I once watched a buck approach a tree. Thinking he was about to start a rub and provide me with great behavioral footage, I hastily got my camera ready. Instead of rubbing the tree, he started to nibble at a patch of small white mushrooms sprouting from the bark! I was surprised, but I did photograph him in the act, and as it turned out, they were still great behavioral shots. I have been lucky enough to photograph bucks rubbing trees on many occasions, but a buck eating mushrooms was a new image for my files.

When the preferred food choices for northern deer change in the early fall, so do their daily movement patterns. Deer establish ritualistic travel patterns from secretive bedding areas to choice food resources. Throughout the summer months they migrate late in the day to their summer foods, which may be a field rich in succulent plants, or a field of farmed clover. They often remain in their feeding area through the night and return to the security of their day bed area (often a thick stand of evergreens back in the woods) early in the morning. In autumn they travel to new, seasonally superior forage such as apple orchards or stands of oak trees. Their bedding area will likely stay the same, but now they must travel a new route to their fall foods.

Once, while conducting a study on deer distribution patterns in southern Ontario, I was able to learn the daily movement patterns of certain bucks. I enjoyed watching a beautiful ten-point buck who crossed a forest access road every morning at 6:30 a.m. through the months of July and August. By early September the buck was gone. Day after day I surveyed the same area at the same time, and he never showed. Eventually, about two weeks later, I rediscovered him; he was no longer migrating from the lush marshy area to the forest as he had in the summer. Now he was migrating out from the far side of the forest to a field that housed about a dozen old, abandoned apple trees. He had changed his movement pattern to benefit from the highest-quality food resources available at that time.

During autumn it is essential that all deer build fat stores, but this is most important with the bucks. Since mature bucks consume very little during the rutting period, they must put on enough pounds during the early fall and pre-rut season to help them through the snowy season. In

fact, bucks are so active during the rut that they can lose over twenty percent of their body weight before winter sets in. Quality autumn food resources are a key factor in helping deer populations through harsh winters.

Despite all the myth and lore suggesting that the first heavy frosts of autumn "jump-start" the whitetail breeding season, it is actually the amount of daylight that dictates when the first does come into heat. White-tailed deer have a tiny pineal gland in their brain, which measures daylight and enables them to respond to changes in the photoperiod (amount of sunlight available in a day). As the days become shorter this gland increases its secretion of the hormone melatonin, which controls a second gland located at the base of the brain, the pituitary gland. The pituitary gland serves in the manufacturing of the repro-

By late October the first does will come into heat, drawing the attention of the bucks.

This large-bodied buck is rubbing a thick tree that suits his size, and probably reflects his status in the dominance hierarchy.

ductive hormones. One of these reproductive hormones controls the production of testosterone from the buck's testes; another controls the production of estrogen from the doe's ovaries. Both testosterone and estrogen are the key hormones that form the basis for the mating drive of white-tailed deer.

As the morning frosts of late October cover the ground, the days will have become short enough to trigger the first does to come into heat, or "estrus." With their newly polished antlers, bucks now devote almost all of their energy to seeking out and courting receptive females.

By this time the bucks will have created many signposts to alert receptive does to their presence. Signposts are strategically placed rubs and scrapes. A rub is an exposed section of tree or shrub where a buck has removed the bark with the rough base of his antlers. Bucks will rub vegetation of various sizes, from the smallest saplings to trees with a

When a rub is complete it will carry the scent of its maker, serving as a signpost to other deer. This tree must have been rubbed by a very impressive buck!

girth exceeding twenty-five inches. The largest bucks almost always tear the bark from the widest trees. Occasionally, a young buck will rub a large tree without much to show for his efforts. When selecting a tree, bucks generally choose young trees about two inches in diameter. These preferred rubbing trees usually have no lower branches and have smooth bark. Bucks prefer rubbing a fresh, unmarked tree to re-rubbing one that they or another buck have already visited. When stripping the bark, a buck deposits scent from his forehead glands onto the exposed cambium, or underlayer, of the tree. This odorous secretion signals the buck's presence with a warning to other bucks that they are on his turf. It also serves to tell the does which buck has been working the area.

When making a rub, the buck will often stop and smell or lick the exposed underlayer of the tree, probably to assess the amount of scent he has deposited. A mature buck will spend about two to three minutes rubbing a tree. Most rubs can be found where different habitats converge; where woods and fields meet is a favorite location. Rubbed trees can also be found along travel corridors between a buck's bedding and feeding area.

Rubs act not only as an olfactory cue but also as a visual cue. With the bark removed, the whitish cambium can be seen from a considerable distance, attracting the attention of other deer. A buck will rub the side of the tree that he approaches. When

A doe approaches a fresh buck scrape. If she finds its pungent scent appealing, she may urinate on or near it and possibly wait in the area for the buck's return.

Bucks make scrapes under an over-hanging branch, referred to as a licking branch. The buck will stretch his swollen neck and nibble at the tip of the branch, leaving his scent. These licking branches are irresistible to bucks.

you stand and look at a rub, you are facing the direction the buck was travelling. In an area that is heavily rubbed, the rubbed trees may serve as markers along the buck's trail and show the direction of his travel.

Dominant, prime-aged males make considerably more rubs than the younger, subordinate males. These mature bucks can easily be distinguished by their large swollen necks. Glands increase the girth of the necks of these more dominant males. The muscles in their necks also strengthen at this time due to all the rubbing activity. This linebacker look gives these individuals an intimidating appearance, reminding younger males that these more mature bucks are in the fittest years of their lives and should not be toyed with.

As the mating season continues the number of rubbed trees in a buck's home range steadily increases. As the peak of the rut draws near the size of the rubs will also increase. Deer typically have a home range of about one square mile, but during the rut, bucks may travel much farther in search of receptive does.

Rutting behavior such as licking branches and making scrapes will continue until the rut is over. In the north snow may have fallen by this time.

Scrapes are another means for rutting bucks to mark their territory and announce their presence to other deer. Scrapes are made when bucks use their front hooves to paw the leaves and detritus off an area of soil of about two square feet. While constructing a scrape a buck deposits another smell signal from a gland located between his toes. After clearing the ground the buck will urinate on the site, leaving his musky odor. While urinating on a scrape the buck will rub his hind legs together, passing urine over his tarsal glands. The tarsal glands are located about halfway up a buck's back legs. These glands are on the inside of the legs and serve to carry the buck's pungent odor to other deer. This strong smell probably serves as an attractant to does and a warning to other bucks. The tuft of hair around these glands becomes darkly stained by the time the rut starts.

The location chosen for a scrape is almost always under an overhanging branch. When making the scrape the buck will take advantage of this other opportunity to disperse his scent by stretching his swollen neck and nibbling at the tip of the overhanging twig; this is referred to as branch licking. This provides another way for bucks to assert their presence in a given area. Licking branches are typically about five feet above the ground. If the only suitable branch is higher, bucks will stand on their hind feet to reach it. I have seen some licking branches as high as seven feet.

Most scrapes are constructed by dominant bucks, and some sites may be used repeatedly for several mating seasons. Deer that walk past the signpost will know which buck's range they are trespassing on. Young bucks will visit scrape sites and often nibble the licking branch, likely as a form of communication rather than an expression of rutting dominance. These less mature bucks rarely scrape the ground, or freshen any already existing scrapes. I have never seen a young buck scrape the ground in the presence of a mature buck. If one did I would wonder whether it was bravery or stupidity. Mature bucks are quite protective of their scraping sites; during the peak of the rut they will chase the younger bucks away from their scrapes. If a mature buck is not around, a younger buck may paw at the ground, practicing for the years ahead. If there is a shortage of mature bucks in an area, younger males will assume some of the more dominant behaviors. These young, adolescent males are usually suppressed by the visual, chemical, and auditory threats of the older, breeding bucks.

If a doe in heat wanders by a scrape and finds the mature buck's pungent scent appealing, she may urinate on or near the site, and possibly wait for him to return. The urine of does that are in heat smells different from that of other does. This difference is due to the presence of sex pheromones, hormone substances that stimulate mating behavior in male deer. Bucks can easily detect these

As does come into heat, bucks like this prime-aged ten-pointer become very active. They spend all their time searching out and courting potential mates.

pheromones with their keen noses. A buck will repeatedly check the scrapes that are in the core of his range and if he encounters the perfume of a receptive female he will immediately pursue her.

This is when the real action starts. Locating, courting, and protecting a hot doe from other bucks is a frantic affair because the doe will only be in heat for about one day! This leaves little time to smell the flowers.

To effectively track the doe that has left the irresistible fragrance, the buck will continue to test her scent trail for the presence of sex pheromones. By curling his upper lip and inhaling air toward a sensory organ in his upper oral region, a buck can verify the doe's state of estrus.

By the time the buck locates the doe there is a significant chance that she will already have a suitor. When this happens the two bucks will size

When following a doe that may be in heat, a buck will sniff the ground where she has urinated and lift his head and lip curl.

up each other and most often recognize their different rankings in the dominance hierarchy. The smaller, subordinate buck will always surrender the doe to the larger buck. Occasionally, the two bucks are of equal size and neither is prepared to give up the doe. This is when white-tails show how powerful and ruthless they can be. With ears laid back and hair bristling, each buck will attempt to intimidate his opponent. If both bucks persist they will suddenly plow into each other with tremendous force. With antlers smashing together and each buck pushing with all his might, something will eventually give. Usually the weaker buck will break free and run from the victor with only a few cuts and a shattered ego. However, sometimes the results of these fights are much more serious. Bucks are often gored by the antlers of their opponents, or antlers can be rammed and broken. On rare occasions the antlers of the two bucks may become locked, resulting in death. With their racks fused together both bucks will eventually succumb to predators, or starve.

During these battles the old warriors who have passed their prime run a high risk of being killed by a younger, stronger buck. Older bucks that have experienced the success of being the dominant male have a difficult time recognizing their physical deterioration and will often accept the challenge of the new boss buck. This is almost always a mistake; rutting whitetails show no mercy.

These violent fights are exhausting to both animals and cause a considerable energy drain. Some battles are over in about a minute, but others have been known to last considerably longer. Following such a match the fatigued combatants are vulnerable to attack from subordinate males, who may take advantage of the larger bucks in an attempt to elevate their own status in the dominance hierarchy. I have seen many instances in which a younger buck tried to challenge a weakened dominant male. Almost always the dominant buck has seen the younger buck approaching in this unusually bold fashion and turned to face him, sending the smaller buck running. If the dominant male has been weakened to the point where he is unable to face the approaching buck, then his breeding season may end.

During one November evening, just before dark, shortly after I had packed my camera away (of course!), I was watching a nearby hillside when an unexpected fight erupted. There were several deer in this small clearing and since it was too dark to photograph I decided to sit and watch them and enjoy their presence. I turned away for what felt like a second when a sudden smashing sound jerked my head around. A large buck, which had been with a doe, had charged a ten-point buck that had ventured too close and knocked him off his feet! The larger buck had the ten-point buck pinned to the ground. Within a couple of seconds, the ten-point buck was able to wriggle free and run twenty yards up the hillside. This display of speed and power had already impressed me, but it wasn't over. A third buck, a respectable, wide eight-pointer, had ventured onto the scene, and he and the large buck both started to walk toward the extremely angry ten-pointer. With ears flattened and hair standing on end, these two bucks converged on the ten-pointer at forty-five-degree angles. The ten-pointer looked back and forth, deciding whether to charge one of the approaching bucks, all the while emitting several aggressive snort–wheeze vocalizations. As the two bucks neared, the ten-pointer—realizing the bullying two-on-one situation—turned and quickly trotted up the hill into the forest. At this point the large buck and the wide eight-pointer were so worked up that they turned and charged into one another! They had an all-out fight for the next minute or so; then the eight-pointer broke free and ran from the larger buck, once again leaving the larger buck with the doe. The speed and ruthlessness of this event left me in awe. Once you've seen whitetail bucks fight you won't ever forget it; it is an incredible sight!

The number of intense fights that may occur in a given area is largely dependent on the buck-to-doe ratio—that is, the number of bucks compared to does. The more adult bucks present per

doe, the more fights. In surveys that I have conducted in two separate areas containing wild whitetail populations, the buck-to-doe ratio has been approximately one to four. This figure can vary substantially from one area to another. In rare locales where the number of mature bucks is close to that of does, the number of violent, strenuous fights will increase.

Once a buck has a hot doe to himself he will approach her to check her receptiveness. When a mature buck approaches a doe that he thinks may be in heat, he will utter a vocalization referred to as the tending grunt. This vocalization is a series of short, quiet, deep grunts followed by a louder, pursuing grunt. Usually the doe will run, leading the buck on a chase to test his fitness and to visually assess his suitability, all the while avoiding close contact until she is ready to copulate.

A young buck is reminded that he has some growing to do and is not yet ready to win sparring matches with older males.

While in pursuit of a hot doe, this buck is about to be challenged by another large buck that has also picked up the irresistible scent.

When two mature bucks fight over a doe, it is a sight to behold!

Unless occupied with mating, bucks will usually investigate any doe that they encounter in the hope that she will be receptive. Until a doe reaches her peak of estrus she remains very shy and will not let any bucks within reach. If a buck pursues a doe that has not achieved her sexual peak, she will do everything she can to avoid him, while giving him some very obvious hints that she is not ready. These hints include fleeing from the buck, bedding down to make it impossible for the buck to mount her, and trying to lose herself in amongst other does. These cues, as well as her neutral scent, signal to the buck that she is not interested, enabling him to go off in search of a more fruitful encounter. A younger, less experienced male is more likely to harass females by continually attempting to make physical contact, even before they are ready to breed. These overzealous males will always have to chase a doe farther than would an older, more experienced buck. Normally a doe will never copulate with such

The pride of any whitetail suitor, a beautiful doe can create havoc in the woods.

an annoying suitor, unless of course all the more mature bucks are preoccupied with other females. While a dominant buck is focused on one doe, subordinate males will take advantage of the opportunity to court and breed with other does that may come into heat at the same time.

A doe in heat will exhibit very clear behavioral cues to signal available bucks. She will hold her tail straight out behind her at a slight angle to the side while trotting with an obvious spring to her step, acting in a very frisky manner. She will also squat and urinate frequently, as often as several times a minute, filling the air with her irresistible perfume. This quickly draws the attention of nearby bucks!

Sometimes there will be three or four bucks in pursuit of a hot doe. The most dominant buck is first, closest to the doe, followed by subordinate bucks, each one at a respectful distance from the one ahead. The pursuing bucks hope that the larger buck will be distracted away from the doe.

Once, while photographing in a large field, I watched while a doe crossed, followed by an impressive ten-point buck, followed by a handsome, slightly smaller eight-pointer, followed by a smaller eight-pointer, followed by a younger, small six-pointer, each about forty yards apart. It was great just to sit there and photograph these different bucks as they crossed the field! In a sense it was almost humorous, this careful game of follow the leader.

During the peak of the rut the home range of a whitetail buck may increase significantly. Usually a buck's home range is about one square mile, sometimes less. During the peak of the rut, while searching for does, bucks may leave their home range and travel five or more miles away. However, if there are enough does in the buck's home area to keep him occupied, and he isn't displaced by a larger buck, then he will stay in his usual area.

Does will most often stay in their home range during the rut. In their typical home range of one square mile or less, does will usually be searched out by eager bucks. If buck numbers are low, a doe may have to travel from her home range to mate, returning after she has been bred.

Does will choose to mate with the male that has the largest body and rack. When the doe is ready she will stand and be bred. The pair will remain together for about one day (twenty-four to thirty-six hours) and will copulate several times. While the buck is tending the doe he will stay quite close to her, never losing sight. While she feeds he will guard her, about ten yards away (rarely feeding himself); when she moves to another area he goes with her; when she beds, he beds by her. When she's up feeding or on the move he will routinely approach her about four or five times an hour to check if she's ready to copulate again. The buck stays with the doe while she is in estrus to prevent other bucks from breeding her, protecting his genetic investment. When the doe is no longer in heat the buck will leave in search of another receptive female. A mature buck has the reproductive potential to breed several dozen does.

Mature does that are in optimum health come into estrus first, followed by younger does. When food resources are less than ideal adult does may become nutritionally stressed, resulting in the delay of their peak period of estrus. Most healthy does in the northern United States and Canada experience their twenty-four hours of peak estrus during the second week of November.

Even though the peak period of mating activity for northern whitetails occurs in November, there is a phenomenon known as the "second rut" which occurs in early to mid-December. The timing of this less dramatic "second rut" is influenced by two factors. First, female fawns born the previous June may enter their first period of estrus during mid-December. Up to forty percent of the female fawns may be bred during their first year of life. However, female fawns that are nutritionally stressed due to poor-quality food resources may not reach puberty until their second autumn. High herd densities may trigger this delay by reducing the quantity of the best foods. Male fawns do not become sexually mature enough to breed until their second autumn.

The second factor causing the "second rut" to

This large buck has picked up the scent of a doe in heat. Following her trail has brought him within sight of her.

This buck has just chased off an eight-point buck and is now herding the doe away from the scene so that he may court and hopefully mate with her.

occur is, if a doe is not successfully bred during her first heat, she will cycle into heat again approximately twenty-eight days later. This once again causes a flurry of activity among the bucks.

Adult does may cycle through as many as three heat cycles in the north and possibly five or six in the south. However, it is quite rare for a doe to require more than one heat cycle to be bred; in most areas there are enough mature bucks to accommodate all the does. In the northern part of the whitetail range, where winter plays a significant role in the lives of deer, a doe will only be successful at producing and raising fawns if they are conceived during her first or second heat. If the doe hasn't been bred by her second heat cycle then there is virtually no chance that she will

Once together, a buck and doe will mate several times over a period of a day or so.

produce fawns that will be large enough to survive the following winter. Although several months away from the fawning season, the timing of the rut in the north is critical to the survival of the fawns. Whitetails have a gestation period of about 200 days. This enables most fawns to be born in June, when food resources are at an optimum. When a doe comes into heat too early in the fall she will give birth too early in the spring, when there is little or no vegetation for the newborn fawn to use as camouflage, and the temperature may be too cold. If a doe is bred after the rut (late December or January) she will give birth during late July, leaving too little time for the fawns to prepare for winter.

In the southern U.S. and Mexico, where seasonal variations are less significant, the rut is stretched out over a longer period. This makes for a slightly less intense peak to the rut. Without the pressure of an oncoming winter and the pressure to have the next generation of fawns grown and prepared for their first winter, deer in the south breed later in the year (often starting in mid- to late December) and breed for a longer period. The peak rut may last for more than a month, and breeding activity may span three months.

Without a snowy winter to dramatically alter their environment, one might ask, why don't whitetails in the southern U.S. breed all year round? Unlike some of the smaller mammals that

A large, dominant buck pursues a receptive doe, patiently waiting for her to stand and copulate with him. Their bond during this time is called the "tending bond."

inhabit the southern, warmer climates, deer still have a ritualized mating season. The key reason for mating during a specific time of year is the importance of the timing of fawning. Even in the south there are seasonal variations, such as drought. In principle, deer have evolved to breed at the time of year that enables fawns to be born when temperatures and food are at an optimum. However, whitetails found near the equator have been observed producing antlers and mating throughout the year.

Another reason that deer rut during a specific time of year, even in the south, is that whitetail bucks have geared their mating strategies around their antlers. In order for all the bucks to compete in the rut, their annually grown antlers must be

With all the does now bred, and the snow depth slowly increasing, this mature, northern whitetail buck may soon move into his winter range.

As the rut winds down and winter sets in, the metabolism of deer will slow down considerably and they will spend much of their time conserving energy in secretive beds.

developed at the same time. This is the only system that male deer have established to ensure that the fittest individuals breed for the benefit of the species.

In the north the rut may come to a sudden halt if an early winter sets in. This may happen in late November or early December in areas like northern Alberta or Saskatchewan. With increasing snow depth and frigid temperatures, optimum, high-energy deer foods become frozen under the snow and ice. Deer now must survive on the woody browse of twigs and evergreens such as cedar and hemlock. These lower-quality foods and the increased heat loss that deer experience in cold temperatures make it no longer feasible for deer to be chasing one another around, expending exorbitant amounts of energy. These winter changes cause a significant drop in deer metabolism, and does stop entering into estrus, switching to a more energy-conserving winter mode.

As the mating season ends and autumn fades to winter, the sex-hormone levels of both male and female deer diminish considerably. These monarchs of the forest once again become shy and passive. Shortly after their antlers are cast, bucks once again begin their preparation for the next mating season, as new antler growth starts to emerge. The does spend their winter conserving energy in anticipation of the considerable task awaiting them in the spring: fawning. As the land hardens in the grip of the northern winter, whitetails prepare for the all-important generation of new life.

FOUR

The Next Generation

The morning sun found the lone doe browsing among some shrubbery in a small field. She was deliberate in her feeding, selecting the rich, new, sprouting plants and leaves. Consuming one leaf at a time, she slowly sated her appetite. Her hunger was considerable at this time of year. She had successfully survived another winter, and her fat stores were spent. The sudden emergence of the green spring growth provided her with a smorgasbord of nutritious forage.

It was late May, and replenishing her own body was not the only physical demand placed upon her. Hidden in the forest was a newborn fawn. The doe now had the task of raising the next generation. She had to nurture the fawn, teach it the ways of nature, and hope for its survival.

With her stomach full, the doe slowly walked back into the woods. Trilliums and violets carpeted the forest floor. These heralds of spring had pushed up through the layer of dead leaves and brought the color of life back to the woods. The leaves on the maples and oaks were flushing and about half the canopy was in. Green abounded; the woods had awakened to another season of plenty.

The doe went about 100 yards back from the field edge. She was nearing the place where she had left her fawn. It had lain silently for the past few hours, resting, growing, and becoming stronger.

By now the sun was high in the sky, bringing warmth and dappling the forest floor with spots of light. The doe bleated. Suddenly, there came a pattering of footsteps as the fawn came running to

(Above) A young fawn bedded in the lush greenery of clover.

(Previous) A whitetail fawn, under the supervision of its mother, grooms along the edge of a summer stream.

its mother. Its little body was quivering with excitement. The fawn quickly touched noses with its mother and then stepped around to her side to nurse. It fed for a few minutes, head bobbing into the doe's udder, encouraging the flow of milk. When the little one had had its fill the doe stepped away. She led her young a short distance through the trilliums and down a small hill, until they came to a stream to drink. After quenching her thirst and allowing the fawn time to explore the cool, clear water, she led him back into the forest. Once they were well back from the creek, the fawn nursed once more; this was followed by a few minutes of licking and grooming from its mother. Clean and content, the fawn slowly left his mother and found a new, secretive bed about thirty yards away. The doe then departed for her bedding spot 100 yards away. There she would rest and chew her cud until it was time to return and feed the fawn again.

Healthy, mature does in the prime of life tend to have twin fawns.

Springtime marks a tremendous shift in the lives of white-tailed deer. This is especially true in the north, where the season of endurance comes to an end and the season of plenty begins. With spring comes an abundance of food resources, allowing deer to recover the weight lost during the long winter months. The fresh shoots and leaves are much more nutritious than anything the white-tails have eaten since the previous autumn. This promise of plentiful foods for the months ahead has caused whitetails to "set their biological clocks," timing their maximum growth and productivity with the return of spring.

Spring is also when the next generation is born. From mid-May to mid-June most healthy, northern whitetail does give birth to their fawns. In the south, where there is not such a significant seasonal change, fawns are born a month or two later. Each doe may produce one, two, three, or even four fawns, depending on her maturity and physical condition. (However, quadruplets are rare.) When food is abundant, and as long as the previous winter wasn't too hard on them, does in the prime of life (four to nine years old) will typically give birth to twins. Young or old does may only produce one fawn. Sometimes a doe will have triplets; however, my research, based on two populations of wild deer, indicated that only five percent of the does gave birth to three fawns. This percentage will vary depending on population density and the quality of available foods. If the number of deer in a given area is high, does are less likely to have triplets than they are in a locale where there are fewer deer and the herd is expanding. In some habitats the quality of foods for whitetails is low in nutritional value, lowering the milk yield of does and making it impossible to raise three young simultaneously.

The severity of the previous winter also influences the number of fawns. Harsh winters with deep snow and limited food resources stress pregnant does. If winter stress becomes severe enough to affect the health of the doe, rendering her unable to properly raise and nourish the coming fawns, she may reabsorb one or more of the fetuses before they are born. This unconscious survival mechanism allows a doe to lower her energy demands during the critical winter months. Without such a mechanism the lives of both the doe and her fawns could be lost. If a doe becomes physically stressed late in her pregnancy, instead of reabsorbing the fetuses, she will likely give birth to stillborn young. This is nature's way of protecting the life of the mother so that she will survive and have the opportunity to breed again next year, hopefully enduring the winter better and producing healthy fawns.

A few days prior to birthing, does will isolate themselves from other deer and move to a choice location to prepare for labor. The more mature,

dominant does will choose the prime birthing locations, lush with vegetation to provide shelter and food. A daughter from the previous year will often establish her fawning territory next to her mother's. This enables the first-time mother to be near her own mother and possibly benefit from her experience. The subordinate does, usually those between three and four years of age, may have to move off and settle for poorer fawning habitat. A doe's dominance status is directly related to age, with the oldest does being the most authoritative.

A doe's birthing zone is usually about ten to twenty acres in size. From the time of delivery until the fawns are a few weeks old, the doe will chase all other deer from this territory, for fear that their odors will possibly draw the attention of passing predators.

Does give birth while standing or lying down. The length of labor varies considerably among individuals, and may last for more than twelve hours. When the fawn has been born the doe will bite off the umbilical cord and consume the after-

Whitetail does will return to their bedded fawns and nurse them about every four hours.

Adult does will hide their young fawns in beds where the lush foliage keeps them concealed from predators.

birth. By eating all of the birthing matter, the doe minimizes the odors left in her territory, reducing the chance of attracting predators. She will then proceed to lick the fawn clean and nurse it for the first time. The doe will stay with her newborn fawn for several hours before taking it to a hidden bedding site and then departing to feed or rest about 100 yards away.

Healthy whitetail fawns usually weigh six or more pounds at birth, with some exceeding ten pounds. The larger subspecies, found in the northern U.S. and Canada, typically have larger offspring than the smaller, southern whitetails.

Whitetails are able to unconsciously control the sex of their offspring. When does are stressed by limited food resources, they tend to produce more male progeny the following spring. When quality foods are plentiful, they're inclined to produce female offspring. In times of high population levels, when resources are stressed, it is better to produce male fawns, because bucks almost always disperse from their mother's home range, decreasing direct competition for food. When food supply is good, warranting immediate population growth in the doe's home range, it is advantageous to produce female fawns that stay in the area and can produce young of their own that very year. The nutritional stress (or lack of stress) experienced by the doe alters some of her hormones, which ultimately influence the sex of the offspring. The varying hormones may alter the uterine environment, favoring one sex over the other.

For the first few weeks following birth the doe will not bed near her fawns. After nursing, the young will wander twenty to thirty yards away from their mother and find their own bedding locations, and the doe will then depart for her bedding site. To control the amount of scent in the area around her fawns, the doe, who has a stronger smell (to predators, not to the human nose) than her newborns, will not bed near them. The doe could escape the jaws of predators, but the young fawns would be caught.

Even though fawns may stand within minutes of birth, their speed and agility are not adequate to outrun most predators until they are about three weeks old. To avoid predation, newborn fawns spend almost all of their time bedded in secretive locations. This is why whitetail fawns have white spots randomly covering their reddish coats—the spots help camouflage the fawns when they're laying motionless. Aside from enhancing the concealment of the nearby vegetation, the spots of newborn fawns also mimic the dappled sunlight on the forest floor. In addition to this handy camouflage, newborn whitetails have virtually no scent.

Siblings are placed in beds away from one another and bed in new areas after each visit from their mother. (Some believe that the fawns choose

This fawn shows how easy it would be to hide amid the lush vegetation of summer. If it had lain down I would never have seen it.

When they are about one month old, fawns will start to accompany their mother on feeding excursions.

their own bedding locations, while others think that in the early stages the mother will choose beds for each fawn.) By spreading out her young, the doe reduces the chance that a predator will discover both of them. For the first couple of weeks the newborns bed eighty or more yards apart. As they get older this distance will diminish, and usually by three to four weeks of age the fawns bed in each other's company.

For the first few weeks following birth the doe will return to her fawns about every four hours to nurse. Each time, the doe will lick the fawn's anal region, encouraging them to defecate. The doe will then eat the droppings, once again minimizing odors. Smelling this region also helps a doe to identify her young. After nursing, the doe will spend a few minutes licking and grooming her little ones. This behavior cleans the fawn and may remove parasites such as wood ticks.

I have unfortunately come across a few situations where hikers have stumbled upon a bedded newborn fawn. They "rescue" the abandoned

As fawns become stronger and more agile, they will spend more time at play, zigzagging back and forth. However, even while frisking they never lose sight of their mother.

fawn and bring it to the nearest Natural Resources office or animal shelter, hoping to save its life. However, the fawn had not been deserted, it was simply bedded down, waiting for the doe to return for its next nursing. Most of these cases end with the fawn being euthanized. Even if it was returned to the woods—if the hikers could remember the exact spot where they found it—it is likely that the doe will have already returned and found no fawn. After searching for it, she will have assumed that it was taken by a predator, and she'll move on with her life. The state of shock that the fawn must be in also decreases its likelihood of survival.

When fawns are three to four weeks old, does reduce the number of nursings per day and start to gradually wean them. Young whitetails start to eat vegetation within a few days of birth, but cannot properly digest plant matter for about two weeks.

At about one month of age fawns spend less time bedded and start to follow their mother on some of her feeding excursions. They are now strong enough to escape most predators. Around this time most people start seeing fawns—especially at dawn and dusk—in feeding areas like fields, or while crossing roads.

When the fawns are about two months old, the mother becomes more sociable and begins to browse in the company of other, related females. Her fawning territory may enlarge to thirty or more acres, and overlap with those of related does. By this time, fawns rely primarily on plant matter for sustenance, and does will reject most nursing attempts.

With the fawns stronger and more active, periods of play seem to erupt spontaneously following nursing. While under the doe's supervision, fawns run frantically back and forth, zigzagging and occasionally bucking. Their mother ignores their playfulness for the most part, but a fawn running at top speed can on occasion startle other adult deer. I have watched a small group of deer feeding peacefully when suddenly all their heads popped up and stared intently into the forest. My first thought was that they had caught the scent of a predator or a stray dog. Instead, the sound of a small running animal could be heard approaching the forest clearing. Moments later, out sprang a frisky, sprinting fawn, bounding through the opening, only to turn sharply and circle back into the woods.

Most often the alarmed adults resume feeding after the fawn has left or calmed down. Sometimes, however, the sound of the running fawn is enough to send the other adults running, tails up, for cover. Adult deer aren't accustomed to hearing another deer running at full tilt unless there is a predator in pursuit, and this is likely their first thought and concern when they hear the approaching fawn. If the adult deer are within sight of the fawn when it starts frisking, they are usually more relaxed and ignore it.

By mid-summer fawns spend virtually all of their time within sight of their mother, even bedding together.

Play serves to strengthen a fawn's muscles and build up its stamina, increasing its chances of outmaneuvering predators. Bouts of playful activity are only evident in fawns, and by autumn they seem to grow out of it and assume the more sedate, mature attitudes of the adult deer.

On several occasions I have observed fawns emerge from hiding and attempt to nurse from the wrong mother. This usually occurs when fawns are aged one month or older and have become increasingly aware and mobile. Normally the doe won't stand for young other than her own: does must conserve their rich milk for their own young. After touching noses with the fawn and not recognizing its scent, she will simply continue walking, leaving the confused youngster to return to its bed. If the estranged fawn persists, the doe may give it a kick with her front hoof to make the message clear. On rare occasion exceptions are made. If a doe has recently lost her own young she

This fawn is accompanying its mother, learning about the succulent foods of summer, and how to always be alert for predators.

A healthy whitetail fawn, alert and aware of its surroundings.

may nurse an unfamiliar fawn. However, this is unusual; most does that lose their fawns continue to reject other young that approach to feed.

An estranged fawn may also succeed in nursing from a doe that is not its mother *while* the doe is feeding her own young. Once while photographing in New York state I was witness to this oddity. I watched as a doe meandered her way through the summer grasses toward the woods. Since it was mid-August, the forest was a thick wall of green leaves, impenetrable by the human eye. When the doe reached the edge an anxious fawn appeared, followed by another, and another, until there were five hungry little ones circling and probably overwhelming her. Obviously, they weren't all hers: there's no way she could have produced enough milk to satisfy the needs of that many growing young. My assumption was that two or three must have been hers and the others likely belonged to another doe in her social group. In late summer, with the fawns partially grown and does slightly more relaxed about their birthing territories, two does may have left their young in the same section of forest. These five were probably bedded close together when the milk-laden doe arrived.

I couldn't believe what I saw. Four of the fawns nursed, their energetic little heads pushing into the doe's udder and actually lifting her hind legs off the ground! They fed for a minute or two before the doe sprang free and walked away, leaving the four fawns satisfied but not feeding the fifth.

If at any time a doe loses her fawns to predators or disease, she will leave her fawning territory and regroup with related females in her larger summer range. None of these does will have fawns. They may not have successfully conceived the previous autumn, or the long winter months may have been too stressful, causing their fetuses to be reabsorbed. Their fawns may have been stillborn or may have succumbed to predation or disease. Sometimes does may abandon healthy newborns if deer densities are too high and there are no satisfactory birthing territories to be found.

Throughout the summer months doe groups tend to inhabit separate ranges from buck groups. Doe groups select habitats that best provide food and cover for the birthing and rearing of fawns. This is usually a mixed habitat, with small clearings and forested sections, and with thick ground cover for fawns. Buck groups prefer areas that simply house nutritious foods that support body and antler growth. They often select a more open habitat and are less concerned with terrain specifics such as ground cover. These seasonal groups rarely mix until autumn.

Despite the segregation of the sexes during the summer months, I have on rare occasion witnessed bucks browsing within close proximity to a doe/fawn group. These sightings occurred in August, when the ranges and protective tendencies of the mothering does had become more elastic and tolerant. I watched while a three-year-old buck fed close to one of the fawns. As the buck neared the fawn he stretched out his neck and smelt the youngster. A moment later he jumped back about three feet, all the while shaking his head, almost as if someone had thrown a glass of water into his face! He immediately approached the surprised fawn a second time, smelled it, and jumped back again. On his third approach he smelled the fawn and then kicked it twice, watched it, and walked away. The fawn lay in shock for a moment, and then it stood up and slowly began to resume feeding. The buck was obviously confused by the fawn and its scent. Why? That's one of the many questions I still have about this dynamic species.

The doe didn't chase the buck away from the fawn because the fawn never uttered a sound. It didn't perceive the buck as a danger, or a threat, and the kicks came so unexpectedly that it never had a chance to bawl for its mother. Typically, however, does are very protective of their young. These seemingly delicate mothers will stand up to and successfully discourage many potential predators. I have watched a doe kick and chase off raccoons on two occasions. Raccoons are not threatening predators to deer, but nonetheless they had obviously come too close to the fawns for the doe's comfort. Defensive does have been known to fend

Through this season of plenty, deer feed on nutrient-rich greenery. A farmer's clover field is no exception.

off coyotes, dogs, sometimes humans, and even black bears. However, I have never heard of a doe trying to stand up to wolves.

Early one August I experienced the display of a defensive doe firsthand. While out surveying the distribution of the different sex/age groups in a population of deer, I heard a fawn bawling about forty yards back in the woods. After listening for several minutes, I decided to investigate. I found the fawn. It had fallen about three feet down into a crevice, between two large sections of rock. I lowered myself to hoist the fawn up, and all the while the doe stomped and snorted and walked around me, trying to draw my attention. I quickly placed the fawn safely on the ground and left, trying to minimize the shock of the event. About thirty minutes later I went back to see if the doe and fawn had gone, and to make sure that the fawn hadn't been injured in the fall. There was no sign of them.

Mature does have been known to act like they themselves are injured in an attempt to lure predators away from their young. A similar strategy is utilized by some birds; for example, a killdeer (a type of plover) will walk while dragging one wing on the ground when trying to distract predators like foxes away from its nest site. When the predator has followed far enough away the bird will jump up and fly off, hoping that the predator has forgotten what had originally caught its attention. Does seem to utilize the same principle when distracting coyotes or bears.

In early fall whitetails molt their reddish summer coats and grow in their thicker, brownish coats for autumn and winter. In the north this molt occurs in early September and lasts for two to three weeks. Adult bucks molt first, followed by adult does without fawns, then yearlings and fawns, and finally does that have young. This is a fawn's first molt, the period when fawns lose their white spots and grow in coats identical to those of the adult deer. Fawns now look like miniature versions of their mothers. Since the fawns are now agile enough to outrun most predators, they no longer require their white spots for camouflage. By now their daily routines are identical to the doe's. The days of lying motionless on the forest floor, hiding from predators, and waiting for their mothers to return to nurse them are long past. Now the fawns feed and bed in synchrony with their mothers.

Whitetails molt their coats for two reasons: camouflage and thermoregulation. As the seasons change and the colors of the fields and forests that make up the whitetail's world go from shades of green to dark brown and eventually back to green, deer must adjust their coloration to remain camouflaged. From spring until the end of summer deer have a bright, reddish pelage. From fall through to spring they have a darker, brownish-gray coat. There are, however, slight variations in pelage color from one geographical region to another. Deer living in the northeastern mixed forests tend to have slightly darker coats than deer found in more open prairie habitats.

A whitetail's coat exhibits another interesting feature. Their underbellies are always white, no matter what season or geographical range. This marking serves as an anti-predator defense mechanism. The white underbelly offsets the deer's shadow cast on the ground, reducing the chance that they'll be seen by predators—the three-dimensionality of the animal's image is reduced, allowing it to blend into its surroundings more.

Thermoregulation is an animal's ability to retain a relatively constant internal temperature in an environment of fluctuating external ones. A deer's coat plays an important role in helping it maintain a steady body temperature. In summer, a whitetail's coat is airy, consisting solely of long, thin guard hairs. This light pelage increases air circulation next to the deer's skin, promoting bodily cooling.

When northern deer molt their summer pelage, the darker coat that grows in consists of higher density, thicker guard hairs. A few weeks after the primary guard hairs appear, when temperatures become frosty, whitetails start to grow their undercoat. This consists of a thick mat of thin, short hairs that greatly increases the insular potential of the coat. The long, hollow guard hairs become coated with oil from the skin glands, making the deer's coat waterproof while the underfur traps the heat.

Whitetails also shed their coats in the spring. In the north this occurs in early May. The timing of both the fall and spring molts are regulated by day length, or photoperiod. Like the many aspects of whitetail biology controlled by seasonally changing light levels, molting is regulated by hormones. As the deer's brain registers the increasing day length in spring (and, conversely, the decreasing day length in fall), internal hormones are produced, triggering molting.

People often misinterpret this shedding appearance as a sign of sick deer, especially in the spring—they claim that the deer look thin or mangy. Admittedly, they are not at their prettiest at this time of year, with tufts of gray winter hair hanging here and there, but their appearance doesn't mean that they're of ill health or have mange. They're simply shedding their thick winter pelage to prevent overheating during the months ahead.

In autumn, one must look at body size to identify fawns that have shed their spots. When they are with their mother, fawns are identifiable by their smaller stature, but body size alone may not always help. When most people observe an antlerless deer on its own, they are often unable to recognize it as a fawn or an adult, sometimes

As autumn approaches, fawns nurse less. By early September, northern deer will molt from their red summer coats to their brown coats of fall. During this molt, fawns will lose their white spots and come to resemble their mother.

assuming that all deer without antlers are does. Aside from body size, an easy trick to identify fawns in fall pelage is to look at their heads: fawns have smaller heads than adult deer. A fawn's skull is shorter than an adult's, with eyes located about halfway between the ears and nose. Yearling whitetails have a slightly longer skull, with eyes located closer to the ears than to the tip of the nose. Finally, adult deer have a wider skull and an even longer snout, and their eyes are located about a third of the skull length away from the ears.

By the time fawns molt to their adult coats, they've been weaned. They may still attempt to nurse, occasionally getting lucky, but the doe will usually walk away, stepping over the fawns and ignoring their request. Once weaned, fawns will remain with the doe and are taught all the behavioral strategies needed to survive the months ahead. The fawns must learn what foods to eat as summer changes to autumn and then winter. In the north, they must also be taught the migration route from their summer territory to their wintering yard.

Fawns remain in the company of their mother until they are one year old. Prior to giving birth to her new fawns, a doe will chase away her yearlings, or young from the previous year. This is not an easy task for the doe: her yearlings are often reluctant to leave her company and start fending for themselves. The doe may have to spend a couple of days repeatedly chasing or even kicking at the yearlings to convince them to leave her birthing territory and allow her to deliver her next generation of fawns.

If female, these solitary yearlings may travel outside their mother's birthing territory and join with female relatives that don't have fawns. Male yearlings will try to join a nearby buck group. Most often they are rejected, however, because of their young age. Sometimes yearlings will band together in a small group for the summer.

By early fall, does with fawns will reunite with their doe social group, which may include the yearling offspring. The yearlings that regroup with their mother are primarily female, but on occasion young males may rejoin the mother's company for a few more weeks.

When the mating season arrives, some of the yearling deer may disperse from their mother's home range and settle in a new location, often several miles away. Virtually all dispersing deer are males. Does tend to be more antagonistic toward their male offspring, which are also threatened by the gestures of the mature, rutting bucks that dominate the maternal range. Female progeny tend to stay and share the same home range with their mother, joining her matriarchal social group.

Fawning is another of the interwoven cycles that occur each year in the world of white-tailed deer. Like the generation of new antlers and the rut, fawning is an essential, well-honed natural process, contributing to the past and future success of these remarkable creatures.

FIVE

Future of the Whitetail

There was a slight breeze in the air and thin wispy clouds stretched across the sky. Inside the forest a handsome ten-point buck was slowly making his way through a stand of beech trees, nearing a field. As he approached the field, though, he turned aside, staying within the cover of the trees. Travelling several yards back from the field edge, the buck walked to a point where the woods ended. He paused momentarily to check for any strange smells or movements. He had to cross twenty yards of ripening hay before regaining the shelter of more trees. His senses picked up nothing unusual, so he cautiously walked across the open space into the neighboring woodlot. Without hesitation he set out through the forest.

After about forty yards the buck came to another clearing. There was a rumbling noise in the distance, and it was getting closer fast. The buck halted and stood still. The noise became louder and louder. The buck had heard this before and knew that if he froze in place the strange thing would pass without bothering him. It happened quickly: the car rushed by while the buck stood motionless just inside the trees. The sound faded, and slowly the woods and clearing became quiet again. The buck still hadn't moved. After a few minutes, he knew the coast was clear, and he jumped across the road and into some thickets. He travelled another 150 yards to a choice feeding area: a farmed field, green with alfalfa. There he would spent the night, intermittently browsing and bedding among the rich vegetation, before venturing back along the same route in the morning.

(Previous)
The secretive whitetail has a promising future. Its superior senses and adaptability will guarantee its continued success.

(Above)
The sight of a deer in the wild will always be captivating.

The majority of white-tailed deer live within close proximity to people. Throughout most of the human-dominated North American continent, people have removed virtually all of the expansive forests that were present prior to the arrival of Europeans. The augmentation of arable lands, required to supply food to the booming human population, has altered most of the whitetail's range. Once again deer have adapted, and for them the human expansion has not been a bad thing. In fact, whitetails have benefitted from the immense changes caused by people. Thanks to our agricultural crops, deer have gained a more predictable, highly nutritious supply of food; and when we cleared the big woods to develop "workable" farmland, we also eliminated most of the whitetail's predators.

In fact, whitetails are the only large, wild mammals that have adapted well to human expansion. Wolves, bears, mountain lions, and bison have all had their ranges shrunken considerably, and rarely share habitat with people. In the case of the wolf, once the whitetail's primary predator, this change is significant; in the lower forty-eight states the wolf is listed as endangered. Deer aren't likely complaining, but maybe they should be; this intelligent predator has shaped the whitetail into the swift-footed herbivore that it is by removing the weak, old, and unlucky members of the species. For thousands, even millions of years, these two species have lived and evolved in a close relationship of predator and prey. Today many whitetails no longer hear the howls of wolves echoing above the trees. The long-term evolution of deer may proceed rather differently over the following centuries, in the absence of wolves.

In the areas where wild lands are large enough, predators still exist, roaming, hunting deer, and having young. Throughout much of Canada and in a few pockets of the lower forty-eight United States, wolves still have enough habitat to elude people and live their private, wild lives. Wolves prey on deer throughout the year, effectively removing old, sick, or injured animals. They most often hunt in a pack, although a lone individual can easily bring down a deer... if it catches one.

In fact, this is where the advantage of pack hunting becomes evident. A lone wolf must stalk, chase, and bring down its prey all on its own, whereas a pack can share these responsibilities, increasing their odds of success. (Packs vary in size, but most often they have seven or fewer members.) A pack of wolves doesn't simply run after a deer in one cluster, hoping to catch it; whitetails are too fast for this technique to work reliably. Instead, wolves use cooperative strategy to outwit their prey. When a pack encounters signs of fresh deer, the wolves disperse. Wolves live and hunt within their pack range (some ranges cover hundreds of square miles), and within this range they seem to know every rock, tree, and shrub... or at the very least the lay of the land. Whitetails also spend most of their lives in a home range (usually about one square mile), and they learn every nook and cranny of this habitat. When trying to outsmart their prey wolves benefit from their knowledge of the terrain within their range, while whitetails benefit from their knowledge of their home range. When fleeing, deer choose the quickest routes that won't trap them in a dead end. When the wolves disperse they send a couple of members of the pack out to the side, to flank the deer. Sometimes one or more of the flanking wolves will quietly circle ahead of the deer, *before* the rest of the pack starts the chase. In this way, wolves often take deer by surprise. The deer knows that it's being pursued by the wolves, but it doesn't know that one or more wolves lie ahead, waiting for the pursuing members of the pack to drive the deer close enough for a kill.

Wolves are not the only predators. Both grizzly and black bears will, if given the opportunity, feed on deer. Grizzly bears inhabit the northwestern United States and western Canada, often preferring the seclusion and higher elevations of the Rocky Mountains. Here the whitetail's range does overlap

Deer must always be alert, especially when on the move. Searching for food, or in this case does, increases their visibility to predators.

with the grizzly's, but only minimally, considering the total scope of the whitetail's distribution in North America. Black bears can be found in the northern United States and throughout much of Canada. These shy, forest-dwelling bears share more habitat with white-tailed deer than grizzlies, and are more apt to occasionally feed on very young, dying, or dead deer. Unlike wolves, bears are omnivores. Their diet is made up of plant matter as well as meat, but primarily they feed on vegetation.

Bears are opportunists. If they encounter an injured or sick deer they may attack it. Virtually all deer taken by bears are fawns. Newborn fawns, lying motionless in the forest, may be consumed by a passing bear that discovers them. Bears are also more likely to feed on deer that are already dead, having been killed by other predators, succumbed to disease, or been hit by a passing automobile.

Other predators of the whitetail include coyotes, mountain lions, bobcats, lynx, alligators (in the south), and even golden and bald eagles. Throughout most of the United States and in the southern-

Other than humans, wolves are the primary predators for controlling whitetail populations. In some parts of the northern United States and Canada, wolves still roam and prey on deer.

In areas where black bears and deer overlap, the black bear may be a predator. Despite being more of a herbivore than a carnivore, black bears are opportunists, and if they encounter a newborn whitetail fawn they will prey on it.

most parts of Canada, such as southern Ontario, coyotes also prey on deer. The wily coyote has managed to survive in these limited ranges, close to people. Coyotes will prey on deer of all ages, but are rarely able to take down healthy adults. Wolves are considerably more effective at capturing whitetails than their smaller cousin; coyotes don't kill as many deer as wolves do, and when they're the sole predator, they're less able to keep highly reproductive deer populations under control.

Mountain lions, also known as cougars or panthers, will perch above deer travel routes and pounce on whitetails as they pass underneath. The ranges of these powerful cats have been reduced by people and now only overlap with that of whitetails in and around the Rocky Mountains. In this western habitat, mountain lions tend to prey more often on mule deer than whitetails. In Florida there is also a remnant population of these tawny cats, and here white-tailed deer may be a more common food source. However, due to the expanding human population, the Florida panther is now on the endangered species list. Since these southern cats are few in number, they probably aren't serving as a major predatory threat to the deer population.

Lynx are a northern cat, found in the boreal forests of Canada and in some areas of the northern United States. These long-legged, grayish-brown cats may prey on the northernmost populations of white-tailed deer. Lynx are most effective at capturing deer in winter, when the snow is deep and the lynx's large furred feet allow it to run well on the surface.

Bobcats and eagles tend to prey only on whitetail fawns. However, these smaller meat-eaters will eagerly feed on the carcasses of dead deer.

Wherever whitetails live close to people, another predator appears: stray dogs, which on rare occasions kill deer. However, they do it differently than the other, wild predators, attacking and harassing deer but often taking a considerable length of time to kill them, if they do at all. Stray dogs are more often an indirect predator: relatively few deer are caught by dogs, but the exertion of frequent and extended chases (especially during the winter) can cause deer to exhaust their energy reserves. This drain may threaten the health of the fetuses carried by pregnant does, affecting the fecundity of the herd, or may even endanger the adults themselves.

Today the most abundant predators of white-tailed deer are humans. This deep-rooted relationship was established long before the arrival of the first Europeans. Native peoples across North America relied on whitetails for food and clothing, and even used their bones for tools. Not long after their arrival, European settlers also discovered the importance of deer. Whitetails were a fundamental source of meat for pioneering

Throughout North America the primary predator of the whitetail has become humans. Hunters serve an important role in managing and controlling deer numbers.

peoples. In the early to mid-1900s, most North Americans lived in rural settings, and the majority of them hunted. The hunting tradition has been carried through several generations, and today millions of people still consume venison across North America.

Hunters also serve a very important ecological role: by harvesting whitetails, they keep deer populations at levels that can be supported by the available habitat. Whitetail densities are continually monitored by wildlife researchers and biologists. By observing the health of the habitat as well as annually monitoring the success rate of hunters, and the age and sex of the deer harvested, wildlife biologists can gauge the health of the herd. They then decide on the number of hunting licenses to be made available for the following year.

In most areas, eligible hunters can purchase licenses that allow them to harvest bucks. In some areas the license automatically permits the hunter to remove a deer of either sex. Where buck-only laws apply, antlerless deer tags (allowing hunters to remove bucks or does) may be made available if deer populations increase and biologists want them reduced. By offering antlerless licenses to a selected number of hunters (often via lottery), biologists increase the harvesting of does, which is the most effective way to reduce herd size. Since every mature buck has the reproductive ability to breed with as many as several dozen does each year, virtually all of the bucks would have to be removed to affect population growth. This task would be almost impossible, given that the number of bucks in an area is unknown, and is further complicated by the elusiveness of white-tails. On the other hand, if the does are reduced in number, the number of fawns born the following spring will also be reduced.

The number of deer allowed per hunter also varies from region to region. In some areas each hunter is allowed one buck per season (with one season per year), and in areas where deer populations are higher, each hunter may take up to two deer per day (in the allotted season). Mind you, not all hunters successfully harvest their allowable quota. When hunter success is high, indicating a healthy whitetail population, biologists may keep the number of available licenses the same, or they may recommend an increase. If hunter success is low, indicating a possible decline in deer numbers, biologists may reduce the number of tags, allowing the herd to increase. The general goal is to keep the number of deer at a level sustainable by the habitat.

Aside from their role in keeping whitetail numbers in balance with their environment, hunters have played a significant part in wildlife conservation. In fact, much of the conserved forest lands that we have today are due to the efforts of ethical hunters and their organizations. Whether privately owned or set aside by government, these reserves conserve the whitetail habitat. In order for there to be deer to hunt, there must be woods for the deer to live in. By protecting these woods, hunters and their organizations also preserve the homes of the many other forest-dwelling animals and plants.

Through the money they spend each year, hunters also provide a substantial financial boost to the economy, and many of those funds go directly toward supporting wildlife management. Before embarking on their annual hunting trips, hunters must purchase a hunting license (the fees for which go directly toward wildlife research and conservation); special hunting attire (which may be the mandatory blaze-orange clothing required during gun seasons, or camouflage outfits for bow hunting, or both); firearms and ammunition; camping gear; and fuel. In the United States there is an excise tax on firearms and ammunition, which also helps fund wildlife conservation efforts. When one multiplies these expenses by the millions of active hunters, the money spent becomes considerable.

Despite the efforts of ethical hunters, wildlife managers, and law-enforcement officers across North America, the unlawful poaching of deer still occurs. There are a small number of people across this continent who shoot deer out of season, or at night, because of poverty; their family may need the meat to survive. More often, though, poaching is committed by deviant criminals who shoot deer

In some urban areas where predators like wolves have long been eradicated, and hunting is limited or non-existent due to legislation, deer populations can proliferate.

for their meat or antlers, with no regard to regulated seasons. Often these poachers are connected with illegal markets, where they sell the meat and antlers. Sometimes poachers simply shoot deer for senseless fun, or they remove the antlers alone and abandon the carcass. Deer deserve more respect than this—*any* part of nature deserves more respect than this. We must honor nature and harvest the animals and plants that it provides with reverence and conservation in mind.

Throughout virtually all of North America, deer numbers are kept in balance with the land and its resources by predators, hunters, and sometimes, in the north, bitterly cold, long winters. However, there are some exceptions. In small areas where hunting is not allowed and the natural predators of deer have been eliminated, overpopulation of whitetails results. This usually occurs adjacent to or within suburbs of large cities. At first, people ignore the growing deer herd, enjoying the frequent sightings, watching deer from their kitchen windows. Within a couple of years, though, a "browse line" becomes evident throughout the forests and on preferred shrubbery in the fields. A "browse line" is a visible horizontal line that occurs after deer have eaten all the greenery off the branches within their reach. By now, local naturalists will usually notice that the deer have started to browse on plants that they wouldn't normally eat. The opportunistic deer will next start to visit vegetable and flower gardens throughout the community that borders the over-browsed woods. Collisions between deer and vehicles also increase.

At this point the overabundance of the deer becomes political. What to do? Most people decide that they'd like to see the population decreased to a more acceptable number, one that the local habitat could sustain. How to do it? This often takes quite a bit of deliberation and frequently heated debate. We have to accept the consequences of our actions; in areas where we have removed their predators, we have to control deer numbers to prevent overpopulation. Typically, there are three options: birth-control drugs, relocation, and culling.

The first option is rarely effective. There are three ways to administer contraceptives: with a dart, via subcutaneous implants, and orally. The difficulty with darting is knowing which does have already been treated and which have not. In one area that I know of, researchers simultaneously shot the does with a contraceptive dart *and* a paint pellet. Unfortunately, this color marker was only visible for a couple of days, after which the same doe could run the risk of being darted a second time—this inefficiency would greatly increase the time and expense involved in the project. The problem with subcutaneous implants is that the does have to be captured to put the implants in place; this is very time-consuming and costly. The third method of birth control, oral administration, has not been widely accepted because of the risk of humans wrongfully consuming the drug. When considering contraceptives the bottom line is cost. To treat enough does in a population takes a tremendous amount of time, and when you consider that a hundred or more does may have to be given the drug, the cost skyrockets. Also, contraceptives may have to be re-administered every few years. The question arises: Who is going to pay?

Contraceptives have another negative effect on deer populations. When given a birth-control drug, does will continue to cycle through periods of estrus for several months. This can be severely detrimental to bucks. They will spend an extended amount of time chasing does, burning all of their winter fat stores, and are then more likely to starve or succumb to the elements.

Relocation is another suggestion often presented when considering the problem of overpopulated deer.

When looked at closely, however, the cost of tranquilizing and relocating whitetails is even higher than the cost of administering contraceptives. In addition, many deer will die during transportation, or even when released into their new habitat. The shock of the event may be too much for many of them.

The final option makes the most sense. Culling, or shooting a percentage of the population, quickly and effectively removes excess deer. Sometimes people suggest hiring a "sharpshooter" to do the job, but this usually adds up to a few hundred dollars per deer. Willing hunters will harvest the deer at virtually no cost. This solution has proven to be the most widely accepted. The local government may establish an application system or a lottery system to select the required number of hunters. Since the deer are often in woods and fields close to suburban residences, they are usually hunted with bow and arrow. This method of population control is the most efficient at reducing deer numbers, and another bonus is that the venison may be consumed.

Automobiles are another aspect of the modern interaction between whitetails and humans. Collisions with fast-moving vehicles have become one of the most significant mortality factors for white-tailed deer. In Michigan alone, over 65,000 deer were hit in 1997. Most of these accidents occur at night, when deer become confused in the headlights of an oncoming car. With the number of paved highways criss-crossing our counties, provinces, and states, collisions with wildlife are a common occurrence. These accidents are usually fatal to deer and may also seriously injure the driver or passengers of the automobile.

Fences are another human-related cause of mortality for whitetails. Throughout their range white-tailed deer often have to jump fences that have been erected by people. Sometimes deer that are weak or unlucky simply misjudge the wire and get caught up in a fence. If a deer's hind legs get hung-up over the fence it is unlikely that it will be able to free itself. Blood loss from lacerations, or wandering predators, usually end the lives of these unfortunate deer.

In the north, some people have taken to feeding deer during the harsh winter months. Unknowingly, these people may do more harm than good. Whitetails, being ruminants (cud-chewing animals), need a few weeks of *gradual* exposure to a new food source to develop the necessary micro-organisms in their stomachs for proper digestion. If at any time during the winter humans begin to supplement deer diets out of concern over visible weight loss, whitetails may fill up on the food provided (usually bales of hay or piles of corn) while being unable to digest it. The introduced foods may just sit in their stomachs, and the deer may starve to death on a full stomach. Hay will simply block the deer's stomach, restricting the passage of digestible foods. Corn is especially dangerous to deer in the

winter because they are not able to handle high doses of starch, which lead to bacterial infection and death.

Several years ago, in southern Ontario, I came across an old-timer who had routinely fed deer for several years. Through the summer and fall he had been dedicated, dispersing corn twice a week. He had stopped feeding when the rut started, because the deer had dispersed and were more preoccupied with their mating rituals. It was late March before he returned to the home range of the deer. He brought several bags of corn, and in their hungry state the deer readily consumed them. The old man returned about a week later to drop off some more corn, and what he found broke his heart. When walking through the area he found the carcasses of eight mature bucks and several does. Upon further investigation it was revealed that they had died with stomachs full of corn. The old man had come to know these particular deer over the past several years and was thoroughly disheartened by the event.

Throughout North America, as more roads are constructed, and faster cars drive these smooth asphalt lanes, more and more deer will be hit.

This was not the first time that human intervention in nature had ended undesirably. However, people here and there across the northern United States and Canada still continue to provide food to wild deer throughout the winter months. When people feed deer in winter it is best to provide them with their natural browse: instead of suddenly dropping off ten bales of hay, people should go and cut more of what the deer are naturally eating (often white cedar) and lay it near their wintering yard. Providing deer with more of the browse that they're presently consuming eliminates the risk of their not being able to digest it. For those who insist on feeding deer foods like hay or corn, it is important that these foods be provided early in autumn and be available *continuously* through fall and winter. This allows for the proper development and maintenance of microbes in the whitetails' rumen (cud-chewing stomachs). In early autumn there are several preferred food resources available to deer, and their rumens house a greater variety of micro-organisms than they do in mid- to late winter. With several preferred foods available, deer are less likely to fill up on the supplemental food. They may even be already feeding on local corn or crops, and may have already developed the required microbes.

In the event of a winter crisis, the Ontario government recommends feeding deer a fifty–fifty mixture of corn and oats. This is suggested when snow depths and lack of available browse become severe, threatening the majority of a deer population.

Of course, the easiest and safest option is to just let wild deer be, and allow them to survive the winter on their own. Winter has always been a critical factor in the lives of northern whitetails, and the deer that continue to inhabit and even expand in these northern latitudes will carry on, enduring this season of hardship.

The future of white-tailed deer looks good, with the exception of two subspecies. The smallest of the whitetails, the Key deer, and the Columbian whitetail (*O. v. leucurus*), are both listed as endangered. The Key deer are found on the islands of the Florida Keys and the Columbian whitetail is found on the west coast, in Washington and Oregon. Both of these subspecies are threatened by the loss of their habitat. This is especially true in the Florida Keys: with a limited amount of land on the islands and increasing human development and tourist activity, Key deer are being squeezed out. Today these deer have a portion of the island designated as a refuge, in hopes of their survival.

Despite the steady expansion of humans, whitetails continue to inhabit most of North America. Their adaptability and wariness have ensured their survival for millions of years and will continue to do so in the future. With sound wildlife management, whitetails and the tranquil forests that they call home will always be around to connect us with nature.

(Left) Adaptable, wary, and secretive, the whitetail has a promising future.

SIX

Photographing Whitetails

The boy walked slowly in the footsteps of his father. His dad had told him that they were many miles from the nearest telephone. The forest they were entering stretched continuously for hundreds of miles. The boy's great-grandfather had grown up in these woods, in a remote home miles from anywhere. His great-grandfather had been raised to live off the land, respecting nature and sharing the woods with the many wild creatures.

Almost a century later the boy and his father were walking through the same woods. The frost-laden leaves crunched under their feet. This was the first time the boy had been taken into the big woods, sharing space with wild creatures like bears, wolves, and deer. Wide-eyed, the youngster followed his dad to their watch. This was the year that his dad had chosen to invite him into an age-old tradition, hunting. Once they arrived at a knoll in the hardwoods, the boy's father told him to walk over the rise and watch for deer on the other side. The boy left his father and walked another forty yards before settling himself on a stump, out of sight from his dad.

The boy sat still. Hours passed while he studied the woods, looking at the different kinds and textures of trees, watching as the last of the autumn leaves broke from their branches and fell to the forest floor, and being entertained by the squirrels as they hurriedly gathered acorns and other forest nuts. The forest sounds were soothing. A slight breeze blew between the trees and the squirrels continually chattered to each other, and every once in a while a raven would fly over, uttering deep gravelly calls.

(Previous) When photographing whitetails it is important to capture an alert expression.

Through the morning the boy became accustomed to the sounds that were common around him in this part of the woods.

Then he heard something different: a rustling of leaves followed by a twig snapping. The rustling sound grew slightly louder as the boy peered through the trees. Moments later he saw it—a deer! It slowly made its way toward the boy, unknowingly heading straight at him. By the time the deer was about fifteen yards away, the boy didn't know what to do. He didn't want the deer to bump right into him, and he was too small and young to carry a gun. He was along on this trip to observe. The boy lifted his arm; the doe froze, stared for a moment, and then bounded away. The boy stood still, caught in a state of awe.

The boy was me. This was the day that marked the beginning of my deep-rooted love of nature, and of deer. This is my heritage, and if you look back into your family's past, it is likely your heritage too. For millions of people, this tradition of being close to nature, and hunting, is alive and well.

I need woods, trees, plants, the smell of pines, and wild animals to feel whole. I have carved my life around nature, living and raising my family by our woods, marvelling at the wild creatures that live so close to us.

As I grew and the question of what I wanted to do with my life arose, I knew that the only way that I could live contentedly was to be close to wild animals and wild places. I decided to study wildlife biology at university. This program taught me a considerable amount about evolution and the natural cycles within nature, not to mention wildlife management principles. But my favorite times were those spent in the company of wild creatures, particularly large mammals like white-tails. Ultimately I decided that I didn't want a desk job as a biologist; instead, I wanted to spend every possible moment in the woods, being close to and learning about the natural world around me. Nothing can match the knowledge gained and the feeling of connectedness with nature and wild animals that comes with countless hours spent wandering in the woods. I decided that I wanted to be a professional wildlife photographer.

This goal would not be easy to attain. First, I had to save up money to acquire the proper camera equipment, working on the philosophy that the best equipment will produce the best results. I had spent many years sneaking up on wild whitetails, trying to photograph them with inferior equipment, and the results were often frustrating. At first I didn't even have a telephoto lens! I started in my mid-teens; whenever I had a free evening after school I would walk to a local wildlife reserve and spend hours quietly walking through the deer trails, trying to get close to one or more of these wary creatures. Sometimes I would lay and wait in clearings where I expected deer to come and browse. I usually watched from a distance with binoculars, observing and learning their behavior before trying to get closer and take a photo.

Over many years, I was able to collect the necessary cameras and lenses to properly photograph wildlife. For wildlife photography, high-quality durable equipment is a must. I have always used Nikon 35mm cameras and lenses. The smaller, lighter 35mm camera bodies are highly preferred by wildlife photographers over medium- or large-format cameras. Those larger cameras are great for landscapes, but are too bulky to be continually carried over the shoulder while wandering through forests and over rocks pursuing animals like white-tailed deer. Film is also cheaper for 35mm cameras than it is for the larger formats. In wildlife photography you have to act fast to capture the action happening in front of you, possibly shooting many pictures to get one that turns out the way you want. This technique would be too costly with larger-format cameras and their more expensive films.

My first camera was a manual Nikon body that had no frills or computer chips. I believe that to truly understand the components of a camera—and the important elements required to make good pictures—a photographer must understand the basics. I feel that the best way to learn about photography is on a manual camera. Today I use the most modern, high-quality cameras because of their advanced light-metering systems and fast motor drives, and occasionally for action shots I use their autofocus modes. However, I still appreciate and often use their manual modes.

A high-quality professional telephoto lens is probably my most important tool, although it is useless without a steady tripod. I use a tripod all of the time. Lenses suitable to photograph wild animals and birds range in length from 300 to 600mm. My choice is a 500mm. This lens gives me ten times magnification (magnification power of a lens can be calculated by multiplying the lens length by two and dividing by one hundred), and is less bulky than the 600mm. Longer lenses allow you to stay farther away from the animal and still get good photos. People that use smaller telephotos are often trying to get closer and closer to the animal in order to get a good picture. If the photographer is able to stay farther back, however, animals are less likely to feel threatened and more likely to behave naturally. Images of animal behavior are the most difficult to get and are often the most desirable. Examples of good behavioral footage of whitetails may include two bucks fighting, a buck rubbing a tree or chasing a doe, or a doe and her fawn secretively drinking from a summer stream.

Aside from the 500mm, I always carry a smaller zoom lens. My favorite is the 80–200mm. When

This doe and fawn are drinking secretively from a summer stream. Photographs that open a window into the hidden lives of whitetails are especially appealing.

animals allow me to get close enough, this smaller lens, being a zoom, allows me to compose the image without continuously moving myself backward and forward. By simply adjusting the zoom I can compose a photo showing a close-up of the animal and then zoom out to get a completely different perspective of the animal in a natural scenic backdrop, showing more of its habitat.

With smaller lenses some wildlife photographers use a "gun stock" instead of a tripod. A "gun stock" is sometimes an actual gun stock removed from a firearm, while other times it is a piece of wood carved into a similar shape. A camera mount is fixed to the top and a cable release is fed down to the finger position. This technique will work with a 300mm or smaller lens, and lends itself to greater mobility, but in my opinion there is no substitute for a good tripod.

Several important factors contribute to a wildlife photographer's ability to take good pictures. Having the proper equipment is essential, but equally important is understanding animal behavior. Knowing the annual cycles of whitetails is very important when trying to capture the various aspects of their lives on film. For example, if I want to get beautiful, majestic shots of whitetail bucks, I don't pursue them in the spring. In springtime they're molting their winter coats and may appear scruffy, and their antlers are just starting to grow. Come October, however, their antlers are fully developed and their body condition is at its annual peak. Still, autumn is not the only season that finds me out searching for deer. In summer whitetail fawns and bucks in velvety antlers can be seen and photographed—two of the more appealing images of the deer year.

Fawns can be photographed from late June through August. My favorite time to try for them is mid-July to mid-August. During this period they are quite mobile, often accompanying the doe on her feeding excursions, which increases their visibility. Throughout this time they continue to look small and gangly, and are still covered with white spots.

I photograph bucks in velvet from May until they shed in early September. However, most of my images of bucks in velvet are taken in July and August, during the later stages of antler development. By early July the bucks' racks have branched out to form the main beams and tines, and the color of their coats works well with the lush greenery of their surroundings.

Understanding behavior also plays an important role in photographing each individual animal. It is important to be able to interpret an animal's body language and gestures, to recognize whether the animal is relaxed, alert, afraid, or agitated. Agitation is more of a concern when working with a bull elk or moose, which may actually attack the photographer; whitetails are more apt to simply flee.

This buck is a superb representative of the species. The orange-colored shrub in the background adds to the appeal of the image.

Understanding body language also helps when trying to anticipate the animal's next move. For instance, if I am watching a mature buck following a doe, and he stops and lowers his head to smell where she has just walked or urinated, I quickly set up and focus on the buck. I know that when he lifts his head he will start lip curling. This is a great behavior to illustrate the presence of the rut, and if I wasn't able to predict that the buck was going to do that upon lifting his head, I most likely wouldn't get the shot. They only lip curl for a few seconds, making it virtually impossible to get set up and focused on them in time to catch the behavior unless you're already prepared.

Knowing where to find and photograph whitetails is another challenge that requires patience, persistence, and luck. There are many places to collect images of whitetails: national parks, provincial and state parks, private estates, wildlife reserves or conservation areas, enclosures, and areas near urban centers where deer have become abundant and less afraid of people. I have found that my most satisfying encounters with deer have been in national and provincial parks, in large private estates, and on my own acreage. On my property I photograph deer by situating blinds in areas of peak activity. Blinds are concealed vantage points, and may be natural (consisting of foliage or vegetation) or artificial (usually tents made of camouflage netting). These areas are always around favorite food resources, such as my field of corn or clover, the apple orchard, a stand of oaks where acorns fall in autumn, or the cedar swamp where the deer yard up in the winter. Wherever I photograph deer, hunting is limited or not allowed.

With wildlife photography it is often more efficient to spend the time travelling to where animals are more visible and abundant than it is to stay put and search in one local area. By travelling and photographing in several areas it is possible to collect images of many different deer, pictured in various kinds of terrain.

One November several years ago, I embarked on a trip to photograph whitetails. I had travelled for two days and arrived late at night. I was tired from the long drive and decided to go straight to bed. I parked in a clearing very close to where I had seen and photographed deer in the past. The cheapest bed for a travelling photographer is the mattress in the back of his pick-up truck. Of course, my truck had a cap on it to keep me dry and I had a very well insulated sleeping bag for warmth. On this night I was very thankful for that, because the temperature dropped below freezing. When I woke up, at dawn, all the windows were iced over with frost and I couldn't see outside. I dressed quickly, very quickly. Anyone who has camped out on a frosty night and then crawled out of their warm sleeping bag into the

Different types of light allow for good pictures. Here an overcast light, which eliminates shadows, works well to show the mood of the weather.

cold, frigid air, only to put on refrigerated clothes, knows exactly how I felt on that November morning.

Once dressed, I opened the back latch of the truck cap and saw an unforgettable sight. Only twenty yards from my truck was a huge buck standing chest deep in frosty grass with three does! They didn't stand for long: when they saw me open the truck latch and pop my head out they were probably as shocked to see me as I was to see them. They bounded off into a nearby section of woods. What a picture that would have been! It will always be in my mind.

However, that wasn't the end of that morning, or that buck. I knew the travel patterns of the deer in this area and knew that these four were heading in from their nighttime feeding area to their daytime bedding location. The section of woods that they ran into was a small island of a woodlot, probably too small for this big buck to want to stay in for the day. On the far side of the woodlot was a field of about two acres in size, which I suspected the buck would cross to gain the shelter of the larger woods beyond. I walked to the field and set up my tripod and camera along the edge of the larger woods, waiting and hoping. Ten minutes later the does emerged and started to cross the field. Not far behind was the buck. As he stepped out I was ready, and I photographed him for a minute or so before he entered the larger bush. Images like these mean the most to me—ones that take me back to a special moment when I was lucky enough to be close to such a majestic animal.

Good photography requires good light. What is good light? It's that soft, warm light found at the beginning or end of a clear day. On sunny days when there are few clouds in the sky the best times to photograph (or more aptly put, the *only* times to photograph) are the few hours following sunrise and the few hours before sunset. Throughout the middle of the day the sunlight is too bright or harsh to create good pictures. The shadows are also too stark throughout the midday hours. In the northern United States and Canada, the days become shorter in fall and winter. The sun is lower in the sky during this time of year, extending the amount of good light, so that appealing photos can be taken until late morning and again earlier in the afternoon.

Another type of appealing light occurs on days when there is a light overcast. Thin, even clouds can create good, soft light, and reduce or even eliminate shadows. With this type of cloud cover the sun is still somewhat visible as a faint, glowing circle behind the clouds. With this light it is best to photograph from mid-morning through to mid-afternoon, when the sun is high in the sky and providing enough light through the clouds. Earlier or later in the day would be too dark. This is the best kind of light for photographing whitetails in their autumn colors.

Photographs depicting specific aspects of whitetail behavior, like rubbing, are the most difficult to acquire, making them the most marketable.

A good quality film is important; select one that will accurately represent the colors and produce a tack-sharp image with fine grain for maximum resolution. Slide films are virtually always the film of choice for professional wildlife photographers. I use Fujichrome Velvia (50 ASA) and Provia (100 ASA). These two films have very fine grain (due to their slow speeds of 50 and 100), and produce vivid natural colors.

Being able to compose images takes practice. Most people have a tendency to place the main subject in the center of the image. This is rarely the best way to compose photographs. Most often the preferred composition is one that places the animal off to one side or the other. For instance, when photographing a buck that is standing broadside to the camera, I set up the picture so that there is less space behind the animal than there is in front of it. I leave a little bit of room behind the deer (taking care not to cut off its behind—unless I'm doing a close-up of its chest and head), but leaving more space in front of the deer shows where the animal is heading. This is more appealing to the eye than seeing where it has been.

If a deer is standing facing me, I swing the camera and lens to a vertical position. When a whitetail is facing me it is much taller than it is wide, and a vertical composition becomes more appealing. I make sure that I don't cut off the deer's feet (unless I am once again doing a close-up of its chest and head) and try to leave some habitat in front of the animal, such as tall grass or a small evergreen. Photographs that show habitat around whitetails are the most appealing because they relay the wildness of their world. In fact, some of my favorite photographs are those that show the deer set slightly back in a natural scene.

When focusing, I always make sure that the animal's eye is sharp and clear, and that there is a glint of light shining from it. The spark of light in its eye brings it to life in a photograph. Natural light will always provide this glint; when it's not visible I wait, as usually only a slightly different angle of the head will catch the light properly.

Depth of field is another important tool for creating good photographs. This is the amount of depth of the picture that appears in focus, and may vary from a couple of inches to infinity. With a shallow depth of field the background of the image becomes faded. Sometimes, when taking a close-up photograph of a deer, I set the lens to a shallow depth of field, like f2.8, so that the woods behind the animal blend from a collage of leaves to a backdrop of green. Depth of field is controlled by the aperture setting, or f-stop, on the lens. The adjustment of the aperture allows for more or less light to pass through the lens and onto the film. When the aperture is opened to a lower f-stop—let's say f2.8—more light will pass through the lens than if the f-stop is set at f11. The lower f-stops

This image depicts a buck in early evening light, drinking cautiously from a stream. The colors of the picture, as well as the buck's reflection, make it a favorite.

allow photographers to continue shooting in the dim light of dawn or dusk. Photographs taken at a lower f-stop will have a shallower depth of field than those taken at a higher f-stop. Therefore, by manipulating the aperture I can influence the amount of detail present in the background of the photo.

Adjusting the aperture affects the shutter speed of the camera. When photographing whitetails I avoid opening the aperture if it slows the shutter speed below 1/30th of a second. At such a low speed I worry about vibration blurring the image. The deer has to be standing very still for a picture to turn out at such a slow shutter speed. To accurately stop a running whitetail I find that I need a shutter speed of 1/500th of a second, or faster.

Photographing whitetails as they change and grow through the seasons and over the years has been, and continues to be, a wonderful adventure. Waking before dawn to hike or drive to where I photograph; feeling the surge of adrenalin as I spot a handsome buck in beautiful light; hurriedly setting up while trying to compose the picture and focus before he moves away; discovering a newborn fawn lying in a field of clover; the thrill of photographing bucks sparring, their antlers clashing, sending wild, echoing sounds through the woods—these are some of the experiences that have made photographing whitetails special to me. For millions of people across this continent whitetails are nature's wild spirits. Seeing them and learning about their lives reminds us what it means to be wild and free.

Selected Bibliography

Bubenik, G.A. (1996). "Morphological Investigations of the Winter Coat of White-tailed Deer: Differences in Skin Glands and Hair Structure of Various Body Regions." *Acta Theriological* 41.

Bubenik, G.A., D. Schams and C. Coenen (1987). "The Effect of Artificial Photoperiodicity and Antiandrogen Treatment on the Antler Growth and Plasma Levels of LH, FSH, Testosterone, Prolactin and Alkalin Phosphate." *Comparative Biochemistry and Physiology*, 83A.

Crompton, L. and J. Bastedo (1998). "Where the Wild Things Are" in *Up Here: Life at the Top of the World* 14(5).

Forsyth, A. (1985). *Mammals of the Canadian Wild.* Camden House Publishing Ltd., Camden East, Ontario.

Gerlach, D., S. Atwater and J. Schnell (1994). *The Wildlife Series: Deer.* Stackpole Books, Mechanicsburg, Pennsylvania.

Mech, L.D. (1970). *The Wolf: The Ecology and Behavior of an Endangered Species.* University of Minnesota Press, Minneapolis, Minnesota.

Ozoga, J.J. and D. Cox (1988). *Whitetail Country.* Northword Press, Minocqua, Wisconsin.

Ozoga, J.J., L.J. Verme and C.J. Bienz (1982). "Parturition Behavior and Territoriality in White-tailed Deer: Impact on Neonatal Mortality." *The Journal of Wildlife Management* 46.

Voigt, D. (1997). *Guidelines for Winter Feeding of Deer in Ontario.* Ontario Ministry of Natural Resources.

Index

Page references to photos are italicized.